AF531334

POPULATION GENETICS

ENCYCLOPAEDIA OF GENETICS

Vol. V

POPULATION GENETICS

By

Dr. Arvind N. Shukla

School of Studies of Zoology & Biotechnology
Vikram University
Ujjain

DISCOVERY PUBLISHING HOUSE PVT. LTD.
NEW DELHI-110 002

First Published-2008

ISBN 978-81-8356-304-8 (Set)

© Author

Published by:

DISCOVERY PUBLISHING HOUSE PVT. LTD.
4831/24, Ansari Road, Prahlad Street,
Darya Ganj, New Delhi-110002 (India)
Phone: 23279245 • Fax: 91-11-23253475
E-mail: dphbooks@rediffmail.com
dphtemp@indiatimes.com

Printed by : Sachin Printers, Delhi

Dedicated to

Prof. Ram Rajesh Mishra

Hon'ble Vice Chancellor

Vikram University, Ujjain [M.P.]

Preface

The present title "*Encyclopaedia of Genetics*" is an exciting, and dynamic branch of science and offers the finest approach to teaching genetics through the integration of the molecular and chemical subdisciplines. It prepares the students to learn to formulate genetic hypothesis and apply critical thinking skill necessary for problem solving, while also gaining a sense of the social and historical context in which genetics has developed. This text also has a completely novel way to illustrate the one or two experiments in each chapter that are rigorously examined according to the scientific method. It starts with the premise that the syllabus for a university course in genetics should reflect the major research issues of the new millennium rather than those topics that were in vogue during the last decades of nineteenth century.

The text covers both the basic and practical aspects of Genetics. Fundamental knowledge is developed within the context of applied relevance. Principles are supplemented with examples. This is done to maintain student interest, which is essential for learning any subject.

In the preparation of this book large number of books and research papers have been consulted. So no authenticity is claimed.

The author expresses his gratitude to Mr. Wasan and staff of M/s Discovery Publishing House for their whole hearted co-operation in the publication of this book.

The author tried hard to be accurate and upto date in statement and realises the impossibility of completely avoiding errors therefore, the author will greatly appreciate having his attention called to any questionable statement.

Author

Preface

CONTENTS

1

Genetics and Statistical Background

The science of *population genetics* deals with Mendel's laws and other genetic principles as they affect entire populations of organisms The organisms may be human beings, animals, plants, or microbes. The populations may be natural, agricultural, or experimental. The environment may be city, farm, field, or Wrest. The habitat may be soil, water, or air. Because of its wide-ranging purview, population genetics cuts across many fields of modern biology. A working knowledge has become essential in genetics, evolutionary biology, systematics, plant breeding, animal breeding, ecology, natural history, forestry, horticulture, conservation, and wildlife management. A basic understanding of population genetics is also useful in medicine, law, biotechnology, molecular biology, cell biology, sociology, and anthropology.

Population genetics also includes the study of the various forces that result in evolutionary changes in species through time. By defining the framework within which evolution takes place, the principles of population genetics are basic to a broad evolutionary perspective on biology. From an experimental point of view, evolution provides a wealth of testable hypotheses for all other branches of biology. Many oddities in biology become comprehensible in the light of evolution: they result from shared ancestry among organisms, and they attest to the unity of life on earth.

Practical applications of population genetics are extensive Many applications, particularly those relevant to human beings, also have important implications in ethics and social policy. Among the applications of population genetics in medicine, agriculture, conservation, and research are.:

1. Genetic counseling of parents and other relatives of patients with hereditary diseases.
2. Genetic mapping and identification of genes for disease susceptibility in human beings, including breast cancer, colon cancer, diabetes, schizophrenia, and so forth.
3. Implications of population screening for carriers of disease genes, confidentiality of results, and maintenance of health insurability.
4. Studies of the heritability of IQ score and its implications for affirmative action, welfare, and other social programs.
5. Statistical interpretation of the significance of matching DNA types found between a suspect and a blood or semen sample from the scene of a crime
6. Design of studies to sample and preserve a record of genetic variation among human populations throughout the world.
7. Improvement in the performance of domesticated animals and crop plants.
8. Organization of mating programs for the preservation of endangered species in zoos and wildlife refuges.
9. Sampling and preservation of germplasms of potentially beneficial plants and animals that may soon vanish from the wild
10. Interpretation of differences in the nucleotide sequences of genes or amino acid sequences of proteins among members of the same or closely related species

The genetic and statistical principles underlying population genetics are for the most part simple and straightforward, but it may be helpful to preface the discussion with a few key definitions and concepts.

Gene Expression and Gene Interaction

Gene is a general term meaning, loosely, the physical entity transmitted from parent to offspring in reproduction that influences hereditary traits. Genes influence human traits such as hair colour, eye colour, skin colour, height, weight, and various aspects of behaviour—although most of these traits are also influenced more or less strongly by environment. Genes also determine the makeup of proteins such as hemoglobin, which carries oxygen in the red blood cells, or insulin, which is important in maintaining glucose balance in the blood. Genes can exist in different forms or states. For example, a gene for hemoglobin may exist in a normal form or in any one of a number of alternative forms that result in hemoglobin molecules that are more or less abnormal. These alternative forms of a gene are called *alleles*.

From a biochemical point of view, a gene corresponds to a region along a molecule of DNA (deoxyribonucleic acid) DNA is the genetic material A molecule of DNA consists of two strands wound around each other in the form of a right-handed helix (the celebrated "double helix") Each strand is a polymer of constituents called nucleotides, of which there are four, conventionally symbolized A, T, G, and C according to the nitrogen-rich base that each contains — either adenine (A), thymine (T), guanine (G), or cystamine (C). The paired strands are held together by weak chemical bonds (hydrogen bonds) that form between A and T at corresponding positions in opposite strands or between G and C at corresponding positions in opposite strands. Wherever one strand contains an A, the other across the way contains a T; and wherever one strand contains a G, the other across the way contains a C. Because of the pairing of complementary bases—A with T and G with C—a double-stranded DNA molecule contains an equal number of A and T nucleotides as well as an equal number of G and C nucleotides DNA molecules can be very long. The DNA molecule in the bacterium *E. coli* is about 4.7 million base pairs, that in the largest chromosome in the fruit fly *Drosophila melanogaster* is about 65 million base pairs, and that in the largest human chromosome is about 230 million base pairs. Physical manipulation of such large molecules is impractical. In order to be studied, they must first be broken into smaller pieces.

Gene Expression

Most genes code for the polypeptide chains that constitute proteins. The code is the sequence of nucleotides along the DNA. In the decoding of the nucleotide sequence in DNA and also in the synthesis of proteins, several types of RNA (ribonucleic acid) are essential RNA is also a polymer of nucleotides, each of which carries a base Three of the bases in RNA (A, C, and C) are the same as those in DNA. The fourth [want (31)i is different. When an RNA strand pairs with a complementary strand of DNA, U in the RNA pairs with A in the DNA. Hence, the base-pairing role of U in RNA is the same as that of T in DNA.

The essentials of gene expression in the cells of higher organisms (eukaryotes). The coding regions of the DNA in a gene, which code for amino acids, are often interrupted by one or more non-coding regions known as *intervening sequences* or *introns*. In the first step in gene expression (transcription), a molecule of RNA is produced that as complementary in base sequence to one of the strands of DNA. Every gene includes a regulatory region (sometimes more than one) that determines when transcription takes place, the types of cells in which it takes place, and the strand that is to be transcribed. Because of the base pairing rules, a DNA sequence—say, 3'-ATCG-5'—results in a complementary RNA sequence—in this example, 5"-UAGC-3'. Note that the DNA and RNA strands each have a polarity or directionality, The terms 5' and 3' refer to the polarity of the strands. The 5' end typically terminates with a free phosphate group and the 3' end typically terminates with a free hydroxyl group (—OH) When two strands of nucleic acid are paired, the polarity of each strand is opposite to that of the other. In the duplex DNA, for example, the left-to-right polarity of one strand is 5'-to-3', whereas the left-to-right polarity of the partner strand is 3'-to-5'. Similarly, in transcription, the template DNA strand has a left-to-right polarity of 3'-to-5', whereas the RNA transcript has the left-to-right polarity of 5'-to-3'. Because of the complementary base pairing between DNA and RNA nucleotides, the base-sequence code in DNA becomes converted into a base-sequence code in RNA, in transcription, the base sequence present in the introns is also faithfully copied into the base sequence of the RNA transcript.

The second step in gene expression in eukaryotes is RNA processing, The beginning and end of the RNA transcript are chemically modified and the introns are removed by splicing (cutting and rejoining). RNA processing results in a molecule called *messenger RNA* (mRNA), in which the (coding regions have been made contiguous. The regions in the original RNA transcript that are retained in the mature mRNA are called *exons*. The central part of the mRNA contains the spliced exons that code for the amino acid sequence of a polypeptide chain. The mRNA also includes exons upstream and downstream from the protein-coding region. The upstream region is the 5' *untranslated region* and the downstream region is the 3' *ontranslated region*.

The final step in gene expression is translation, in which the mRNA molecule combines with ribosomes and other types of RNA molecules in the cytoplasm to produce the final polypeptide. In the coding region of the mRNA, each adjacent group of three nucleotides constitutes a separate coding group or codon that specifies which amino acid is to be incorporated into the polypeptide chain. The ribosome moves along the mRNA in steps of three nucleotides (codon by codon).

As each new codon comes into place, the correct amino acid is brought into line and attached to the end of the growing chain of amino acids. New amino acids are added to the growing chain until a codon specifying "stop" is encountered. At this point synthesis of the chain of amino acids is finished and the polypeptide is released from the ribosome.

In prokaryotes, which includes bacteria and other organisms lacking a nucleus, gene expression is essentially identical to that in eukaryotes except for the absence of RNA processing Genes in prokaryotes do not contain introns and so splicing is unnecessary. In prokaryotes, the original RNA transcript is used immediately as m RNA and translated into a polypeptide. Because there is no separate nucleus, translation in prokaryotes often begins immediately when the 5' end of an RNA transcript comes off the DNA and even before transcription of the 3' end of the same molecule has been completed.

The central role of RNA in gene expression is one of the oddities of biology that makes sense in the light of evolution. That gene expression is configured around RNA is a legacy of the earliest forms of life when RNA molecules served both as carriers of genetic information and as catalytic molecules. The role of RNA as carrier of genetic information was gradually replaced by DNA, and the role of RNA as catalytic molecules was gradually replaced by proteins. At every step along the way, as the RNA world evolved into the DNA world, the role of RNA was indispensable in the processes of information transfer and protein synthesis, and so the RNA intermediates became locked in place.

Genetic Code

The genetic code is the list of all codons showing which amino acid each codon specifies. A few organisms and some cellular organelles, such as mitochondria, use slightly altered codes. The codons are those found in the mRNA. The amino acids are given by three-letter abbreviations as well as by conventional single-letter abbreviations. Codon AUG is the start codon in polypeptide synthesis; it specifies methionine (Met) at the beginning of the polypeptide as well as at internal positions. Three codons are stops that result in termination of polypeptide synthesis: UAA, UAG, and UGA. The genetic code is redundant in that most amino acids are specified by more than one codon. Most of the redundancy is in the third codon position.

A code for an amino acid is *twofold degenerate* if either of two sequences specifies the same amino acid. Twofold degenerate codes have the pattern -Y or -R, where Y stands for the bases in codon positions 1 and 2. The symbol Y stands for any pyrimidine base (either U or C); the symbol R stands for any purine base (either A or G) For example, CAU and CAC both code for histidine (His), fitting the pattern CM; and CAA and CAC both code for glutamine (Gin), fitting the pattern CAR. A code for an amino acid is *fourfold degenerate* if any of four sequences specifies the same amino acid, fourfold degenerate codes have the form N, where N means any nucleotide (U, C, A, or G). For example, GUU, GUC, GUA, and GUG all code for valine (Val), which fits the pattern GUN. Note that the code for isoleucine is *threefold degenerate* and those for leucine, arginine, and senile are each *sixfold degenerate*.

The codons for amino acids are not used randomly in proteins. There are preferred codons for amino acids that differ from one gene to the next and from one organism to another. Codon preferences exist even within *redundancy* classes. In *Drosophila*, for example, among codons for histidine, CAC is used more than CAU in a ratio of about 2 : 1. Similarly, among codons for glutamate, CAG is used more than CAA in a ratio of about 3 : 1. Another example of nonrandom codon usage is the AUA codon for isoleucine, which tends to be avoided in most proteins in most organisms In *Drosophila*, AUL and AUC are used more than AUA in a ratio of about 10 : 1 One evolutionary hypothesis that explains the avoidance of AUA us that, because of the degeneracy of the genetic code, the AUA codon might sometimes be translated as AUG, which codes for methionine. Because methionine is likely to change protein structure radically, the mistranslation would be a costly mistake. Through evolutionary time, one by one, the AUA codons in a messenger RNA become replaced with AUU or AUC, minimizing this type of misincorporation error. This misincorporation hypothesis for AUA codon avoidance has not been tested, but it is testable.

Alleles

Alternative alleles of a gene differ in their sequence of nucleotides. For example, where one allele has a T-A base pair in the DNA, another may have a C-G base pair at the same position Because of redundancy in the code, not all nucleotide substitutions result in a replacement of one amino acid for another. For example, if a mutation at the third position in the second codon (asterisk) changes one pyrimidine into the other, the new codon still codes for histidine On the other hand, some nucleotide substitutions at the third position do result in amino acid replacements. For example, if the third position in the second codon changes from a pyrimidine to a purine, the codon changes from one for histidine to one for glutamine. Most nucleotide substitutions at codon positions one and two result in amino acid replacements.

Not all alleles differ by a mere nucleotide substitution. Relative to the typical or wildtype allele, some alleles may have a deletion of a number of nucleotide pairs or an insertion into the DNA molecule, The number of nucleotides deleted or inserted may be small (as few as one nucleotide pair) or large. Some insertions are thousands of nucleotide pairs in size. Many large insertions result from the activity of *transposable elements*, which are specialized sequences of DNA able to replicate and insert at novel positions virtually anywhere in the DNA of the organism in which they are present Alleles also may differ in the number of copies of short sequences present in tandem arrays in the DNA. For example, near many genes in human beings are tandem copies of &nucleotides, such as 5'-CACACACA ... -3'. Such a repeating sequence is symbolized as (5'-CA-3')n The number of copies (n) of the &nucleotide repeat often range from fewer than ten to hundreds, and the number of copies may differ dramatically from one allele to the next. Some alleles even differ from wildtype in having an inversion of the nucleotide sequence in a region of DNA.

Genotype and Phenotype

Within a living cell, genes are arranged in linear order along microscopic threadlike bodies called *chromosomes*. A typical chromosome may contain several thousand genes The position of a gene along a chromosome is called the *locus* of the gene. In most higher organisms, each cell contains two copies of each type of chromosome. Such organisms, in which the chromosomes are present in pairs, are said to be *diploid*. In each pair of chromosomes, one member is inherited from the mother through the egg and the other is inherited from the father through the sperm At every locus, therefore, diploid organisms contain two alleles, one each at corresponding positions in the maternal and paternal chromosomes. If the two alleles at a locus are chemically identical (in the sense of having the same nucleotide sequence along the DNA), the organism is said to be *homozygous* at the locus under consideration; if the two alleles at a locus are chemically different, the organism is said to be *heterozygous* at the locus. The term *gene* is a general term usually used in the sense of *locus*.

Geneticists make a fundamental distinction between the genetic constitution of an organism and the physical or biochemical attributes of the organism. The genetic constitution of an organism is called the *genotype*; genotype thus refers to the particular alleles present in an organism at all loci that affect the trait in question. For example, if a trait is influenced by two genes, each with two alleles, then there are nine possible genotypes, as follows:

AA ; *BB*	*AA* ; *Bb*	*AA* ; *bb*
Aa ; *BB*	*Aa* ; *Bb*	*Aa* ; *bb*
aa ; *BB*	*aa* ; *Bb*	*aa* ; *bb*

where *A* and *a* refer to the alleles of the first gene and B and *b* refer to the alleles of the second gene. In some cases when the genes are *linked* (located in the same chromosome), it is sometimes necessary to distinguish between the genotypes AB/*ab* and *Ab/aB,* in which case there are ten possible genotypes.

In contrast to genotype, the physical expression of a genotype is called the *phenotype*. Examples of phenotypes include hair colour, eye colour, height, weight, number of kernels on an ear of corn, number of eggs laid by a hen, and round versus wrinkled pea seeds The distinction between the genetic constitution of an organism (genotype) and the physical or biochemical attributes of the organism (phenotype) is particularly important in cases in which the environment can affect the trait; in such cases, two organisms with the same genotype can nevertheless have different phenotypes because of differences in the environment. Conversely, two organisms with the same phenotype can have different genotypes.

Dominance and Gene Interaction

Whether each genotype has a single, unique expression of the trait depends on the manner in which the alleles of a gene interact in development For the alleles of one gene, *dominance* refers to the concealment of the presence of one allele by the strong phenotypic effects of another. For example, with two alleles there are three possible genotypes:

AA *Aa* *aa*

Several types of dominance are distinguished and exemplified in the following examples:

1. *Complete dominance: A* is completely dominant to *a* if the phenotypes of *AA* and *Aa* cannot be distinguished.
2. *Incomplete dominance: A* shows incomplete dominance with respect to *a* if the phenotype of *Aa* is intermediate between that of *AA*

and that of *aa*. This situation is also referred to *aa* partial dominance or intermediate dominance. When the phenotype can he measured on a quantitative scale, for example, the number of kernels on an ear of corn, and the phenotype of *Aa* is exactly the average between that of *AA* and that of *aa,* then the alleles are said to be additive alleles and the type of dominance is sometimes called *semidominance*.

3. *Codominance: A* and *a* are codominant if the products of both alleles can be detected in *Aa* heterozygotes. Many alleles are codominant at the level of their protein products because two different forms of the polypeptide, encoded by A and *a,* can be detected in heterozygotes. At the level of the DNA sequences, all alleles differing in DNA sequence are codominant.

It is important to note that dominance is not a characteristic of alleles so much as a characteristic of the manner in which the phenotype is examined. An allele may show complete dominance if the phenotype is examined in one way, no dominance if examined in another, and codominance if examined in still another. For example, the allele for round pea seeds *W* studied by Gregor Mendel is completely dominant to that for wrinkled seeds *w* when the phenotype "round" versus "wrinkled" is examined. The genetic defect in wrinkled seeds is the absence of an enzyme needed for the synthesis of a branched-chain form of starch. Microscopic examination reveals subtle differences in the form of the starch grains in seeds of the three genotypes: W seeds contain large, well-rounded starch grains, retain water and shrink uniformly as they ripen, so the seeds do not become wrinkled, seeds lack the branched-chain starch and are irregular in shape because the ripening seeds lose water more rapidly and shrink unevenly. However, heterozygous *Ww* seeds have starch grains that are intermediate in shape even though the seeds shrink uniformly and show no wrinkling. Therefore, at the level of the starch grains, there is incomplete dominance of *W* and *w* because the starch grains in the heterozygotes are intermediate between the two homozygotes. Furthermore, the difference in DNA sequence between *W* and *w* can readily be detected with modern methods, so that *W* and *w* are codominant at the level of DNA sequence.

For traits affected by more than one gene, the relation between genotype and phenotype depends not only on the degree of dominance of the alleles of each gene but also on the type of interaction between the genes in development. For example, suppose that the trait in question is degree of pigmentation and that pigmentation is determined by two alleles of each of two genes, say, *A*, *a* and *B*, *b*. Suppose further that the total amount of pigment in an organism results from the total number of *A* and B alleles present, each of which adds a single unit of pigmentation to the phenotype. Then, there are only five possible levels of pigmentation (0 through 4) and genotypes *aa BB, Aa Bb,* and *AA bb* all have the same phenotype. Because each uppercase allele adds the same quantity to the total phenotype, the type of gene interaction is said to be *additive*.

Segregation and Recombination

The essential mechanism of inheritance was established by Gregor Mendel (1822-1884) in experiments with garden peas carried out in the years 1856 to 1863 in a small garden plot next to the monastery in which behaved. Mendel showed that the alleles of each gene segregate from one another in the formation of reproductive cells or gametes. Because of segregation, heterozygous genotypes form equal numbers of gametes containing each allele. Furthermore, because gametes unite at random in fertilization, the following are the results of simple Mendelian segregation:

AA × *AA* makings produce all *AA* progeny.
AA × Aa matings produce 1/2 *AA* and 1/2 *Aa* progeny.
AA × *aa* matings produce all *Aa* progeny.
Aa × *Aa* matings produce 1/4 *AA*, 1/2 *Aa*, and 1/4 *aa* progeny.
Aa × *aa* matings produce 1/2 *Aa* and 1/2 *aa* progeny
aa × *aa* matings produce all *aa* progeny.

The physical basis of Mendelian segregation is that the maternal and paternal pairs of chromosomes are separated into different cells in the formation of gametes. Prior to their separation, the maternal and paternal chromosomes associate intimately all along their length and alleles may be interchanged in the process of *recombination*. The interchange of parts takes place after the chromosomes have replicated, and only two of the four chromosome strands participate in any one exchange. Recombination results in the creation of allele combinations different from either parental chromosome. Therefore, a single exchange between parental chromosomes results in two recombinant and two nonrecombinant gametes.

In organisms with an XX–XY chromosomal mechanism of sex determination, Mendelian segregation randomizes the sex ratio at fertilization. In mammals and many other animals, sex is determined by sex chromosomes: males have an X and a Y chromosome, and females have two X chromosomes. In males, the X and Y chromosomes segregate, yielding equal proportions of X-bearing and Y-bearing sperm. If both types of sperm are equally able to fertilize eggs, then random union of sperm with eggs yields 1/2 XX (female) and 1/2 XY (male) chromosome constitutions.

Probability in Population Genetics

The basic concepts of probability needed for elementary population genetics are quite straightforward. They will be introduced with the concrete example of genetic segregation, which deals with the progeny of the mating *Aa* × *Aa.* Considerations in probability always begin with an experiment of some kind. The experiment may be either a real experiment or a conceptual experiment. In probability calculations, it is also necessary to define all possible outcomes of the experiment The outcomes are called *elementary outcomes* because they are defined in such a way that, in any repetition of the experiment, one and only one of the elementary outcomes must be realized. For example, if we are interested in the genotypes among the progeny of the mating *Aa*, the possible elementary outcomes for each offspring are either *AA*, *Aa*, or *aa* (Note that, in defining these as the elementary outcomes, we are ignoring the possibility of either *A* or *a* mutating to a novel allele.) To proceed further, we must assign to each elementary outcome a *probability*, a number between 0 and 1 that measures how much confidence we have that the outcome will be realized. The probabilities assigned to the outcomes are based on genetic reasoning, intuition, or experience. One requirement of the assigned probabilities is that the probabilities of all the elementary outcomes must add to 1, this is the mathematical consequence of requiring that one of the elementary outcomes must be realized. For example, if there are three elementary outcomes, and all are equally probable, then each has a probability of 1/3. The probabilities assigned to the elementary outcomes *AA*, *Aa,* and *aa* are 1/4, 1/2 and 1/4 respectively, because these are the relative proportions of the three progeny genotypes expected from Mendelian segregation.

Addition Rule

An outcome of a conceptual experiment is an event. The distinction between an event and an elementary outcome is that an event can include more than one elementary outcome. For example, the event "the offspring has at least one copy of the dominant A allele" consists of two elementary outcomes, namely, genotypes *AA* and *Aa*. This event may be symbollized A-, where the dash indicates that the unspecified allele may be either *A*'or *a*. For events defined in terms of elementary outcomes, the probability of an event equals the sum of the probabilities of the elementary outcomes included in the event. In the present example,

$$\Pr(A\text{-}) = \Pr(AA) + \Pr(Aa) = 1/4 + 1/2 = 3/4$$

More generally, two events are *mutually exclusive* if they cannot be realized simultaneously. The *addition rule* states that, for mutually exclusive events, the probability that either one or the other is realized equals the sum of the probabilities of the separate events.

Multiplication Rule

All possible genotypes of sibships of three offspring from the mating $Aa \times Aa$, with each offspring classified as *A*- versus *aa*. (A *sibship* is a group of brothers and sisters.) The probability of A- in any particular birth is 1/4 , and that of *aa* is 1/4. The probabilities M the right are the overall probabilities for each of the sibships. They are obtained by multiplication of the probability for each birth because successive births are *independent*, which means that the genotype of any birth has no effect on the genotype of any other birth. Because of the independence, among the 3/4, of the sibships with *A*- in the first birth, 3/4 will have A- in the second birth, and among the $3/4 \times 1/4$ of the sibships with *A*- in the first two births, 3/4 will have *A*- in the third birth. Therefore, the overall probability of three A- births is $3/4 \times 3/4 \times 3/4$. The reasoning for the other types of sibships is similar. More generally, the *multiplication rule* states that, whenever two events are independent, the probability of their joint realization is the product of the probabilities of their being realized separately.

Repeated Trials

The sibships are an example of *repeated trials* of a conceptual experiment. Repeated trials are encountered frequently in probability They govern tosses of a coin or dice, deals of cards, successive spins of a roulette wheel, and so forth. Repeated trials are also important in population genetics because successive offspring of a mating are independent events and thus repeated trials. Furthermore, it is apparent that the different birth orders are mutually exclusive any sibship can have one and only one birth order of A- or *aa*. Because the birth orders are mutually exclusive, their probabilities may be combined by the addition rule. Hence, the composite events below have the following probabilities

$$\Pr(\text{two } A\text{- and one } aa) = 9/64 + 9/64 + 9/64 = 27/64$$

$$\Pr(\text{one } A\text{- and two } aa) = 3/64 + 3/64 + 3/64 = 9/64$$

Note that, when the sibships with the same number of A- and *aa* genotypes are combined, the overall probabilities are given by successive terms in the expansion of:

$$(1/4\ A\text{-} + 1/4\ aa)^3 = 1 \times (3/4)^3 \qquad A\text{-}\ A\text{-}\ A\text{-}$$
$$+\ 3 \times (3/4)^2\ (1/4)^1 \qquad A\text{-}\ A\text{-}\ aa$$
$$+\ 3 \times (3/4)^1\ (1/4)^2 \qquad A\text{-}\ aa\ aa$$
$$+\ 1 \times (1/4)^3 \qquad aa\ aa\ aa$$

The coefficients 1 : 3 : 3 : 1 are the number of combinations in which each triad of genotypes can be born: 1 for *A- A- A-*, 3 for *A- A- aa* (because the *aa* genotype can be born either first, second, or third), and so forth. Each power of 3/4 and 1/4 is the probability that any one of the birth orders will be realized, for example, $(3/4)^2(1/4)^2$ is the probability that any sibship with two *A-* and one *aa* genotype will be realized.

In all cases of repeated and independent trials, the overall probabilities are given by analogous expansions. Suppose that any one trial may result in either of two mutually exclusive events, *A* or *B*, and that the probability of event *A* is *p* and that of event *B* is *q* (with $p + q = 1$). Among a total of n independent trials, what is the probability that *A* is realized exactly *r* times and *B* is realized exactly $n - r$ times? By the multiplication rule, any particular combination of *r As* and $n - r$ *Bs* has a probability $p'q^{n-1}$. Deducing the total number of combinations of *t As* and $n - r$ Bs is a little less obvious, but it is given by the coefficient of the term $p'q^{n-1}$ in the expansion of $(p + q)^n$, which equals

$$\frac{n^t}{r!(n-r)} \qquad \text{...(1)}$$

where the exclamation point means the *factorial*, the product of all integers from 1 through the number in question. For example, $n! = 1 \times 2 \times 3 \times \ldots \times n$ For consistency, the number *n* is defined as $0! = 1$.

Equation 1 is often called a *binomial coefficient* because it arises in the expansion of the two terms $(p + q)^n$. To understand the reason why Equation 1 I yields the correct number of combinations of *r As* and $(n - r)$ *Bs*, first consider what the *n* means. It is the total number of ways that any set of *n* objects can he arranged in order. There are *n* ways to choose the first object and, having chosen the first, $n - 1$ ways to choose the second and, having chosen the first two, $n - 2$ ways to choose the third, and so on, yielding $n \times (n - 1) \times (n - 2) \times \ldots \times 1 = n!$ Furthermore, for each arrangement of *n* objects of which r are As and $(n - r)$ are Bs, there are $r!$ ways to arrange the As among themselves and $(n - \text{r})^1$ ways to arrange the Bs among themselves, for a total of $r! \times (n - \text{r})!$ arrangements. Because each of the n^1 combinations of r As and $(\text{n} - r)$ Bs includes $r! \times (n - r)!$ equivalent arrangements of the As and Bs, the total number of different arrangements of r As and $(n - r)$ Bs equals the ratio given in Equation 1.

Equation 1 gives the number of different arrangements of r As and $(n - r)$ Bs. Each arrangement has a probability given by $p^r q^{n-1}$. Therefore, using the addition role, the probability that n repeated trials yields r realizations of A and $(n - r)$ realizations of B equals

$$\frac{n^1}{p^1(n-r)^1} p' q^{n-1} \quad \text{...(2)}$$

As an example of the use of Equation 2, consider the probability that a sibship of 12 offspring from the mating $Aa \times Aa$ perfectly matches the expected Mendelian ratio of 9 $A-$ and 3 aa. In this case, $p = 3/4$, $q = 1/4$, $n = 12$, $r = 9$, and $n - r = 3$. The required probability from Equation 2 is therefore

$$\frac{12^1}{9!3^1}\left(\frac{3}{4}\right)^9\left(\frac{1}{4}\right)^3 = 220 \times 0.0751 \times 0.0156 = 0.258$$

The implication of this calculation is that, whereas the "expected" ratio is 9 $A-$: 3 aa, only a little more than 25% of such sibships actually have the expected distribution.

Phenotypic Diversity and Genetic Variation

One of the universal attributes of natural populations is that organisms differ in phenotype with respect to many traits. Phenotypic diversity in many traits is impressive even with the most casual observation. Among human beings, for example, there is diversity with respect to height, weight, body conformation, hair colour and texture, skin colour, eye colour, and many other physical and psychological attributes or skills. Population genetics must deal with this phenotypic diversity, and especially with that portion of the diversity that is caused by differences in genotype. In particular, the field of population genetics has set for itself the tasks of determining how much genetic variation exists in natural populations and of explaining its origin, maintenance, and evolutionary importance. Genetic variation, in the form of multiple alleles of many genes, exists in most natural populations. In most sexually reproducing populations, no two organisms (barring identical twins or other multiple identical births) can be expected to have the same genotype for all genes. Thus, it becomes important to describe how alleles in natural populations are organized into genotypes—to determine, for example, whether alleles of the same or different genes are associated at random.

Allele Frequencies in Populations

Much of the phenotypic variation in natural populations does not yield simple Mendelian segregation ratios such as 1 : 1 or 3 : 1 in pedigrees. Some differences in phenotype are environmental in origin and so are not expected to show Mendelian segregation. However, simple Mendelian segregation is not usually observed even for traits whose expression is influenced more or less strongly by genetic factors. Although the underlying genetic factors do segregate in pedigrees in Mendelian fashion, the segregation is concealed by several complications First, environmental effects on the trait may be strong enough to mask the genetic segregation. Second, genetic effects on many traits are determined by the joint effects of the alleles of two or more genes, and the segregation of any one gene in a pedigree may be obscured by the segregation of others. On the other hand, some phenotypic diversity in populations does show simple Mendelian segregatior In the snapdragon *Antirrhinum majus*, for example, whether the flower colour is red, pink, or white is determined by the alleles *I* and *i* of a single gene. The genotypes *II, Ii,* and *ii* have red, pink, and while flowers, respectively, an example of incomplete dominance.

Populations containing both the *I* and *i* alleles will include plants whose flowers are red (*II*), pink (*Ii*), or white (*ii*) in proportions determined by the allele frequencies of the *I* end *i* alleles in the population as well as by the manner in which the alleles are united in fertilization By the *allele frequency* of a specified allele, we mean the proportion of all alleles of the gene that are of the specified type. To take a hypothetical example, suppose 400 members of a population were classified as to flower colour and the finding was: 165 red, 190 pink, and 45 white Because the flower colour reveals the genotype, we may infer that the sample of 400 includes 165 *II,* 190 *Ii,* and 45 *ii* genotypes. The observed numbers of *I* and *i* alleles are therefore

$$P\ 2 \times 163 + 190 = 520$$

$$r\ 190 + 2 \times 45 = 280$$

The factors of 2 are included for the homozygous genotypes because each *II* genotype contains two *I* alleles and each *ii* genotype contains two *i* alleles. The total number of alleles in the sample equals 2 × 400 = 800. Therefore, if we let *p* represent the frequency of the *I* allele and *q* represent the frequency of the *r* allele (with $p + q = 1$

because these are the only alleles of the gene in question), then we can estimate p and q from the observations as:

$$\hat{p} = 520/800 = 0.65$$
$$\hat{q} = 280/800 = 0.35$$

Note that, if the *I* and *i* alleles were combined into genotypes at random, the expected frequencies of three genotypes can be calculated from the rule for repeated trials by expanding the binomial $(p\ I + q\ i)^2 = p^2\ II + 2pq\ Ii + q^2\ ii$. Therefore, assuming random combination into genotypes, the expected numbers of the three genotypes are:

$$II\text{: } (0.65)^2 \times 400 = 169$$
$$Ii\text{: } 2 \times 0.65 \times 0.35 = 182$$
$$ii\text{: } (0.35)^2 \times 400 = 49$$

Hence, the observed numbers in this hypothetical population are very close to those expected with random combinations of alleles. The proportions p^2, $2pq$, and q^2 for the three genotypes when two alleles are combined at random constitutes the *Hardy-Weinberg principle*, which is one of the basic principles in population genetics.

Parameters and Estimates

In the discussion of flower colour in snapdragons, we made a subtle distinction between the actual allele frequency of the *I* allele (designated p) and the estimated allele frequency of the *I* allele (symbolized $\hat{p}$). The distinction is necessary whenever an experimenter makes inferences about an entire population from an examination of a random sample from the population. Quantities used in describing entire populations are *parameters*. In the snapdragon example, the parameter of interest is the allele frequency p of *I* in the entire population. Because we only have access to a sample of 400 organisms from the population, the true value of p is unknown, The best we can do is make an estimate of p based on a sample, hoping that the sample is representative of the population as a whole. The estimate obtained from the sample is designated $\hat{p}$ to emphasize that it is an estimate rather than the true valve. Whenever it is necessary to distinguish parameters from their estimates, we use unembellished symbols for parameters (for example p for the unknown frequency of an allele in a specified population) and the same symbol with a circumflex for the estimated value (in this example $\hat{p}$)

Standard Error of an Estimate

The distinction between a parameter and an estimate is important because different samples may yield different values of the estimate for the same reason that different sibships may yield different segregation ratios, namely, chance variation from one repeated trial to the next. The estimation of an allele frequency can be treated as repeated trials by supposing that the alleles are sampled at random, one by one, from a very large population. In the snapdragon example, there are 800 alleles sampled. If the allele frequency of I has the true value $p = 0.65$, then the repeated-trials interpretation implies that all possible outcomes of 800 trials have probabilities given by successive terms in the expansion of $(0.65\, I + 0.35\, i)^{800}$. This is not an expansion that one would to do by hand, but the binomial expression makes evident the underlying random-sampling process that accounts for variation in the estimate of $\hat{p}$ from one sample of 800 alleles to the next.

Unless p is quite close to 0 or quite close to 1, there is a convenient approximation to the binomial expansion $(p\, I + q\, i)^n$, where n is the number of alleles sampled. As n becomes large, the distribution of $\hat{p}$ approaches the familiar bell-shaped curve called the *normal distribution*. The normal distribution features prominently in the analysis of traits determined jointly by multiple genetic and environmental factors and it is discussed in detail that context. For present purposes, it is sufficient to note that the degree to which the values of $\hat{p}$ are clustered around the overall average depends on a quantity called the *standard error*.

$$s = \sqrt{\frac{\hat{p}\hat{q}}{n}} \qquad \text{...(3)}$$

where $\hat{q} = 1 - \hat{p}$. If the sampling and estimation of p were repeated many times using the same population, then the values of $\hat{p}$ would be expected to be clustered symmetrically around p according to the standard error as follows:

1. Approximately 68% of the estimates $\hat{p}$ lie within plus or minus one standard error of p.
2. Approximately 95% of the estimates $\hat{p}$ lie within two standard errors of p.
3. Approximately 997% of the estimates $\hat{p}$ lie within three standard errors of p.

To put the matter in another way, with repeated sampling, 32% of the estimates would be expected to differ from the true value by more than one standard error, 5% by more than two standard errors, and only 0.3% by more than three standard errors.

As an illustration of the variation among repealed estimates of p, the values of p obtained in 100 repetitions of the experiment of sampling 800 alleles from a large population in which the true allele frequency is $p = 0.65$. Each of the 100 samples was created by computer simulation using a random-number generator that yielded a 1 with probability 0.65 and a 0 with probability 0.35. For each sample of 800, therefore, the estimate $\hat{p}$ equals the number of 1s in the sample divided by 800. The distribution of $\hat{p}$ values is more or less bell-shaped but not exactly so because it is based on only 100 samples rather than an infinite number. The overall mean $\hat{p}$ from all 100 samples combined (80,000 observations) equals 0.6492, which is very close to the true value of p. Furthermore, the distribution of the estimates fits the predictions based on the standard error quite well.

To apply Equation 3, note first that $p = 0.65$ with $n = 800$, and so s in Equation 3 equals $\sqrt{[(0.65 \times 0.35)/800]} = 0.017$. Because 68% of the samples are expected to yield values of $\hat{p}$ in the range $p \pm s$, and because the expected distribution is symmetrical, 34 of the values are expected in the range $p - s$ to p (0.633 0.650) and 34 in the range p to $p + s$ (0.650 - 0.667); the actual numbers are 33 in the first interval and 35 in the second. By the same reasoning, 95% of the values should lie in the range $p \pm 2s$, or 47.5% on each side of the mean; because 34% of the values on either side of the mean are in the range $p \pm s$, the implication is that 47.5 - 34 or 13.5 of the values should lie in the range $p - 2s$ to $p - s$ and 13.5% should he in the range $p + s$ to $p + 2s$. These ranges are 0.616–0.633 and 0.667-0.684; the actual number in each interval is 18 and 10, respectively, as against the theoretical 13.5 in each. Likewise, the standard error predicts that 0.3% of the samples will deviate by more than $3s$ from the mean, as compared with the observed 2.

Estimates and their standard errors are often presented as $\hat{p} \pm s$, or 0.65 ± 0 017 in the present example. The 68%, 95%, and 99.7% cutoffs for ±1, ±2, and ±3 standard errors provide one mantle' in which the reliability of an estimate may be interpreted. Estimates may also be presented alternatively in terms of a range called a *confidence interval*, which expresses a degree of confidence that the true value of a parameter lies in some specified interval The most frequently encountered confidence interval is the 95% confidence interval, defined as the interval ($\hat{p} - 2s$, $\hat{p} + 2s$). Because 95% of repeated samples are expected to yield estimates in a range ±2s around the true mean, then 95% of the time the interval ($\hat{p}$ - $2s$) – ($\hat{p} + 2s$) is expected to include the true value of the parameter p. In the snapdragon example with $\hat{p} = 0.65$ and $s = 0\ 017$, the 95% confidence interval is 0.616-0.684.

Models in Population Genetics

Population geneticists must contend with factors such as population size, patterns Of mating, geographical distribution of organisms, mutation, migration, and natural selection Although we wish ultimately to understand the combined effects of all these factors and more, the factors are so numerous and interact in such complex ways that they cannot usually be grasped all at once. Simpler situations are therefore devised, situations in which a few identifiable factors are the most important ones and others can he neglected An intentional simplification of a complex situation is a model. There are several types of models, each designed to eliminate extraneous detail in order to focus attention on the essentials. Some models are experimental An experimental model may consist of a laboratory experiment with population cages of *Drosophila* or growing cultures of bacteria. An experimental model may also consist of observations of natural populations in particular locations or at particular times in which evolutionary forces of interest may be presumed to he present. Models of this type include the study of the origin and spread of insecticide resistance in insects or antibiotic resistance in bacteria.

A model may also be a conceptual simplification Conceptual models have a number of uses. They require a concise statement of presumed mechanisms and interactions; they afford a framework for interpreting observations and setting research priorities; they enable extrapolation into the future or beyond the range of known parameters; and they suggest tests of consistency between theory and observation

A conceptual model may consist of verbal arguments logically linking a chain of hypothesis and deductions. Another type of conceptual model is a computer program that simulates the random component in a process or that calculates the values of changing quantities in a complex system based on prescribed numerical relations. In population genetics, a kind of model frequently encountered is a mathematical model, which is a set of hypotheses that specifies the *mathematical relations* between measured or measurable quantities (the parameters) in a system or process Mathematical models can be extremely useful

1. They express concisely the hypothesized quantitative relationships between parameters.
2. They reveal which parameters ate the most important in a system and thereby suggest critical experiments or observations.

3. They serve as guides to the collection, organization, and interpretation of observed data.
4. They make quantitative predictions about the behaviour of a system that can, within limits, be confirmed or shown to be false.

The validity of any model must be tested by determining whether the hypotheses on which it is based and the predictions that grow out of it are consistent with observations.

A mathematical model is always simpler than the actual situation it is designed to elucidate. A model is supposed to be simple If it is not simpler than the real situation, then it isn't a model. Models are simpler than real situations because many features of real life are intentionally ignored. To include every aspect of a complex system would make a *model* too complex and unwieldy. Construction of a model always requires a compromise between realism and manageability. A completely realistic model is likely to be too complex to handle mathematically, and a model that is mathematically simple may be so unrealistic as to be useless. Ideally, a model should include all essential features of the system and exclude all nonessential ones How good or useful a model is often depends on how closely this ideal is approximated. In short, a model is a sort of metaphor in analogy. Like all analogies, it is valid only within certain limits but,, when pushed beyond these limits, becomes misleading or even absurd

In this book, we are going to take many liberties with mathematical rigor. Our excuse is that the basic ideas of a model are often obscured rather than illuminated by excessive attention to mathematical detail Our authority for the approach is the great physicist Richard Feynman, who wrote in one of his papers:

Mathematicians may he completely repelled by the liberties taken hem The liberties are taken not because the mathematical problems are considered unimportant. On the contrary, to encourage the study of these forms from a mathematical standpoint In the meantime, foist as a poet has a license from the rules of grammar and pronunciation, we should like to ask for "physicists' license" from the rules of mathematics in order to express what we wish to say in as simple a manner as possible.

2

EVOLUTION AND SPECIATION

Populations change, or evolve, through natural selection and the other forces that perturb the Hardy-Weinberg equilibrium. The merger of population genetics theory with the classical Darwinian view of evolution is known as *neo-Darwinism*, or the "new synthesis." In the two previous chapters, we laid the theoretical groundwork for an understanding of the process of evolution in natural populations. In this chapter, we concern ourselves with long-term evolution and speciation.

DARWINIAN EVOLUTION

Charles Darwin was a British naturalist who published his theory of evolution in 1859 in a book entitled *The Origin of Species by Means of Natural Selection, or the Preservation of Favoured Races in the Struggle for Life*. This book provided overwhelming support for evolution as well as a mechanism for the process. Darwin had been greatly influenced by the writings of the Reverend Thomas Malthus, who is best known for his theory that populations increase exponentially, whereas their food supplies increase arithmetically. Malthus, who proposed his theory in *An Essay on the Principle of Population* in 1798, was referring specifically to human populations and was trying to encourage people to reduce their birthrate rather than let their offspring starve to death. Malthus's writings impressed upon Darwin the realization that under limited resources—the usual circumstance in nature—not all organisms survive. In nature, organisms compete for the resources needed to survive.

Darwin sailed aboard the HMS *Beagle*, a ship that circled the world from 1831 to 1836 with the primary purpose of charting the coast of South America. During his travels on the *Beagle*, Darwin amassed great quantities of observations (especially on South America and the Galapagos Islands) that led him to suggest a theory. Darwin proposed that organisms become adapted to their environment by the process of natural selection. In outline, the process works according to the following principles:

1. *Variation is a characteristic of virtually every group of animals and plants*. Darwin saw variation as an inherent property among individuals of all populations.
2. *Every group of organisms overproduces offspring*. Most populations maintain a relatively constant density over time. Thus, every parent, on average, just replaces itself. Therefore, most of the offspring the individuals of a population produce will die before they reproduce. Hence, in every group of organisms, there is an overabundance of young.
3. *Those that do survive and reproduce will pass on their genes in greater proportion*. This step is the cornerstone and the best-known part of Darwin's theory. Among all the organisms competing for a limited array of resources, only the organisms best able to obtain and utilize these resources survive (*survival of the fittest*). If the favourable characteristics of these individuals are inherited, these traits pass on to the next generation. These organisms then have the greatest reproductive success.

Thus, over time, if advantageous mutations arise, or if the environment changes, the characteristics of a population should change through the process of natural selection (directional or disruptive selection). A particularly well-adapted population in a stable environment may maintain its numbers through the forces of stabilizing selection. Nonrandom mating, genetic drift, and migration may also play a role in population differentiation.

Evolution and Speciation

The term *evolution* describes a change in genotypic frequencies, which usually results in a population of individuals better adapted to the environment than their ancestors were. *Speciation* comes in two different forms. (1) It may be the evolution of a population over time until the current population cannot be classified as belonging to the same *species* as the original population. This process is known as *anagenesis*, or *phyletic evolution* (*an* is Latin for without, *genesis* is Latin for birth or creation). (2) Speciation may also be the divergence of a population into two distinct forms (species) that exist simul-

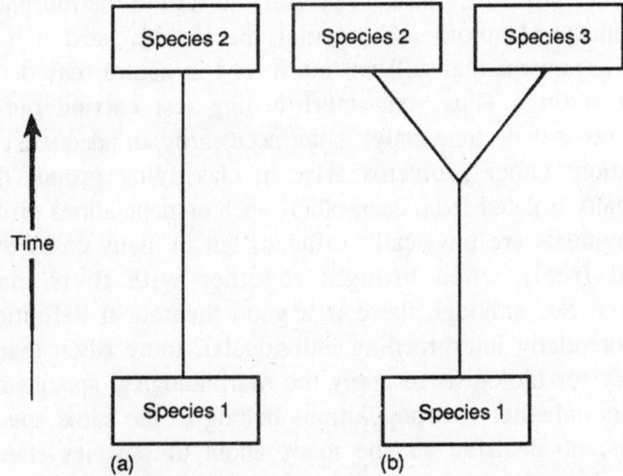

Fig. 2.1. Forms of speciation: (a) anagenesis, (b) cladogenesis.

taneously. This branching process is known as *cladogenesis*. What do we mean by the term *species*?

Before Darwin's time, *typological thinking* prevailed, and a species was defined as a group of organisms that were morphologically similar. All variants were considered imperfections of the model or type. One of Darwin's greatest contributions to modern biological theory was to treat variation as a normal phenomenon in a group of organisms. The modern *biological species concept* groups together as members of the same species organisms that can potentially interbreed. A species, therefore, is a group of organisms that can mate among themselves to produce fertile offspring.

Unfortunately, the definition of species on the basis of interbreeding cannot be used in many places, mostly due to the technical problems of applying it. Taxonomists and paleontologists, who often use nonliving specimens (preserved or fossilized), use the *morphological species concept* as a working definition. Under this concept, two organisms are classified as belonging to the same species if they are morphologically similar. They are classified as belonging to two different species if they are as different as two organisms belonging to two recognized species. Other problems arise for taxonomists since speciation is a dynamic process. For example, isolated subgroups of a population may be in various stages of becoming new species; the rate of successful interbreeding among individuals from these subgroups may range from 0 to 100%. How should the in-betweens be classified? There is no correct answer. It depends on the circumstances.

Still other problems make it necessary to turn to the morphological species concept. Haploid and asexual species are hard to classify. Also, two organisms that will not interbreed in nature may do so in a laboratory setting. Thus, the interbreeding test carried out in the laboratory (as is done frequently) is not necessarily an adequate criterion for speciation. Other problems arise in classifying groups that are geographically isolated from each other, such as populations on islands. These individuals are physically isolated, but in many cases they can interbreed freely when brought together with their mainland counterparts. So, although there is a good theoretical definition of a species (potentially interbreeding individuals), more often than not it is necessary for biologists to apply the morphological species concept to determine whether two populations belong to the same species. In some cases, no decision can be made about the species status of a population. It is clear that a population has evolved, but it is not clear whether it has evolved enough to be called a new species. However, this is more of a problem for taxonomists and evolutionary biologists than for the organisms themselves.

Mechanisms of Cladogenesis

Reproductive isolation

How does one species become two? Basically, *reproductive isolating mechanisms* must evolve to prevent two subpopulations from interbreeding when they come into contact. Reproductive isolating mechanisms are environmental, behavioural, mechanical, and physiological barriers that prevent individuals of two species from producing viable offspring. Following is a modification of the classification system of isolating mechanisms suggested by evolutionary biologist G. L. Stebbins:

1. *Prezygotic mechanisms* prevent fertilization and zygote formation.
   (a) Residential—The populations live in the same region, but occupy different habitats.
   (b) Seasonal or temporal—The populations exist in the same region, but are sexually mature at different times.
   (c) Ethological (in animals only)—The populations are isolated by incompatible premating behaviour.
   (d) Mechanical—Cross-fertilization is prevented or restricted by incompatible differences in reproductive structures.
2. *Postzygotic mechanisms* affect the hybrid zygotes after fertilization has taken place.

(a) F_1 hybrid breakdown—F_1 hybrids are inviable or weak.
(b) Developmental hybrid sterility—Hybrids are sterile because gonads develop abnormally or because meiosis breaks down before it is completed.
(c) Segregational hybrid sterility—Hybrids are sterile because of abnormal distribution to the gametes of whole chromosomes, chromosome segments, or combinations of genes.
(d) F_2 breakdown—F_1 hybrids are normal, vigorous, and fertile, but the F_2 generation contains many weak or sterile individuals.

Allopatric Parapatric, and Sympatric Speciation

Reproductive isolating mechanisms are barriers to *gene flow*, the spread of genes between populations. These isolating mechanisms can evolve in three different ways, each of which defines a different mechanism of speciation. Usually, the mode of speciation is dictated by both the properties of the genetic systems of the organisms and stochastic (random) or accidental events. For example, vertebrates tend to have different speciation modes than phytophagous (plant-feeding) insects.

The appearance of a geographic barrier, such as a river or mountain, through the range of a species physically isolates populations of the species. Physical isolation can also occur if migrants cross a particular barrier and begin a new population (founder effect). The physically isolated populations can then evolve independently. If reproductive isolating mechanisms evolve, then two distinct species are formed, and if they come together in the future, they remain distinct species. Speciation that occurs because reproductive isolating mechanisms evolve during physical separation of the populations is called *allopatric speciation*. As evolutionary biologist Guy Bush pointed out, "Although examples in nature are difficult to substantiate ... it [allopatric speciation] has been convincingly demonstrated in frogs ... and lizards."

Reproductive isolating mechanisms usually originate incidentally to the speciation process. That is, they arise incidentally during the process of evolution in isolated populations rather than being selected for. When isolated populations come together again, incomplete isolating mechanisms may allow hybrids to form. If the hybrids are normal and viable and can freely interbreed with individuals of each parent population, then no speciation has taken place. However, if the hybrids are at a disadvantage, natural selection may favour stronger isolating mechanisms. In this case, organisms that mate with individuals from the other population leave fewer offspring. The result is a more effective barrier to hybridization. Regions in which previously isolated populations come into contact and produce hybrids are called *hybrid zones*.

Until recently, evolutionary biologists believed that allopatric speciation was the general rule. Many now believe that two other modes of speciation may occur frequently in certain groups of organisms. *Parapatric speciation* occurs when a population of a species that occupies a large range enters a new niche or habitat. Although no physical barrier arises, the new niche acts as a barrier to gene flow between the population in the new niche and the rest of the species. Here again, reproductive isolating mechanisms evolve to produce two species where there was only one before. Parapatric speciation is believed to have occurred often in relatively nonvaglle animals such as snails, flightless grasshoppers, and annual plants. *Sympatric speciation* occurs when a polymorphism, which is the occurrence of alternative phenotypes in the same population, arises within an interbreeding population before a shift to a new niche. This mode of speciation may be common in parasites and phytophagous insects. For example, if a

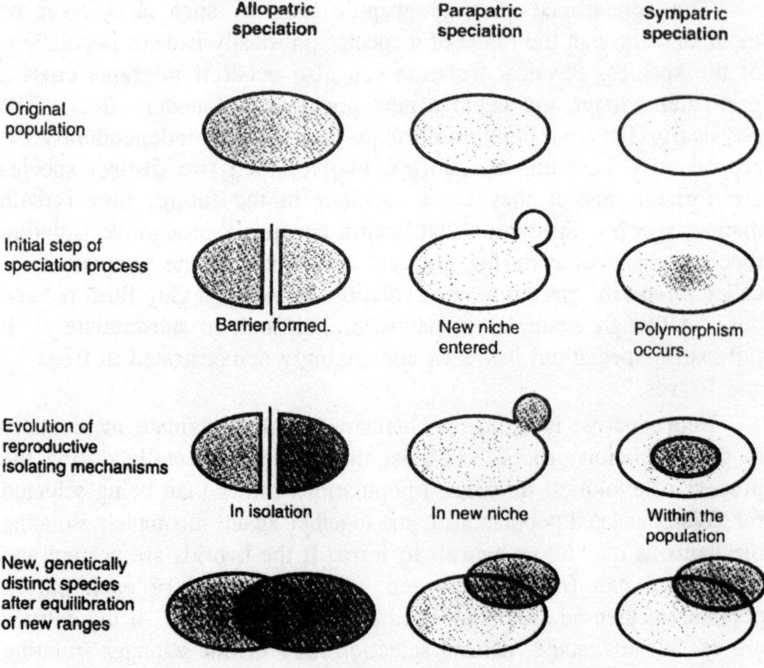

Fig. 2.2. The three general mechanisms of speciation.

polymorphism arises within a parasitic species that allows an individual with a certain genotype to adapt to a new host, this genotype may be the forerunner of a new species. If the parasite not only feeds on the new host but also mates on the new host, a barrier to gene flow arises, although the parasite may be surrounded by other members of its species with the original genotype. Sympatric speciation can thus occur in the middle of a species range rather than at the edges.

An example of incipient sympatric speciation has been seen recently in host races of the apple maggot fly (*Rhagoletis pomonella*) in North America. This fly was found originally only on hawthorn plants. However, in the nineteenth century, it spread as a pest to newly introduced apple trees. In fact, races are now known on pear and cherry trees and on rose bushes. These races have developed genetic, behavioural, and ecological differences from the original hawthorn-dwelling parent. Evolutionary biologists view this as an opportunity to observe sympatric speciation as it occurs.

Another form of sympatric speciation occurs when cytogenetic changes take place that result in "instantaneous speciation." These cytogenetic changes include polyploidy and translocations. For example, if polyploid offspring cannot produce fertile hybrids with individuals from a parent population, then the polyploid is reproductively isolated. This mechanism is much more common in plants because they can exist vegetatively despite odd ploidy and they usually do not have chromosomal sex-determining mechanisms, which are especially vulnerable to ploidy problems.

The end result of cladogenesis is the divergence of a homogeneous population into two or more species. One of the classic examples of cladogenesis appears in the ground finches of the Galapagos Islands. These birds are very well studied not only because they present a striking case of speciation, but also because Darwin studied them and was strongly influenced by them in his views.

An original flock of finches somehow reached the Galapagos Archipelago from South America, 700 miles away, and with time spread to the various islands of the Galapagos Archipelago. Given the limited ability of the birds to get from island to island, allopatric speciation took place. On each island, the finch population evolved reproductive isolating mechanisms while evolving to fill certain niches not already filled on the islands. For example, in South America, no finches have evolved to be like woodpeckers because many woodpecker species already live there. But the Galapagos Islands, being isolated from

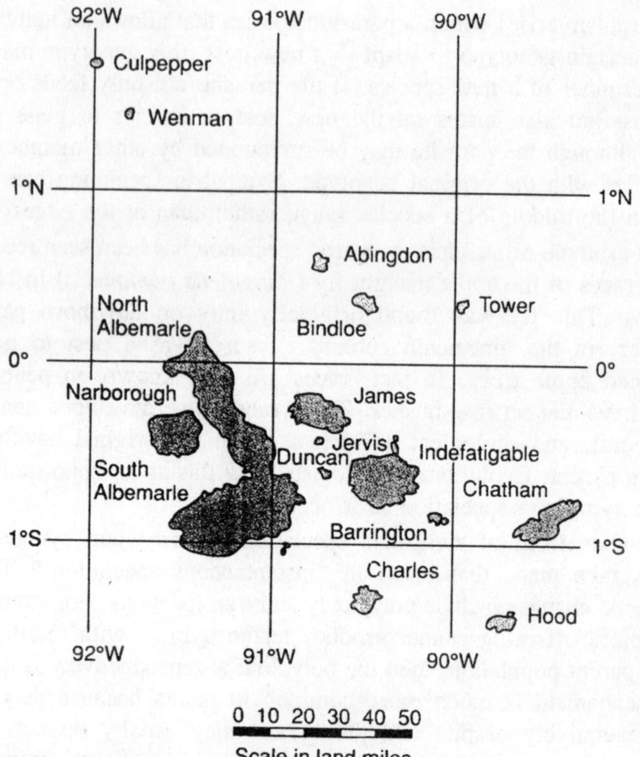

Fig. 2.3. The Galapagos archipelago is located about 700 miles west of Ecuador.

South America, have what is called a *depauperate fauna*, a fauna lacking many species found on the mainland. The islands lacked woodpeckers, and a very useful food resource for birds—insects beneath the bark of trees—was going unused. Finches that could make use of this resource would be at an advantage and would thus be favoured by natural selection. On one island, a finch did evolve to use this food resource. The woodpecker finch acts like a woodpecker by inserting cactus needles into holes in dead trees to extract insects. Darwin wrote: "Seeing this gradation and diversity of structure in one small, intimately related group of birds, one might really fancy that from an original paucity of birds in this archipelago, one species had been taken and modified for different ends."

Phyletic Gradualism Versus Punctuated Equilibrium

Darwin visualized cladogenesis as a gradual process, which we refer to as *phyletic gradualism*. However, an alternative view arose in

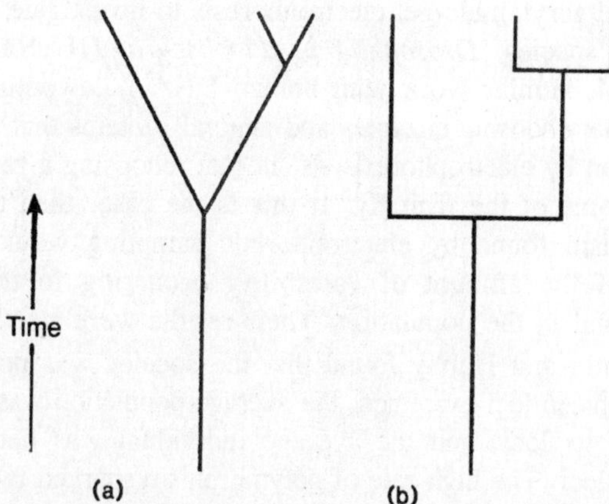

Fig. 2.4. Diagrammatic interpretation of cladogenesis. (a) Phyletic gradualism is depicted as a gradual divergence over time. (b) Punctuated equilibrium is depicted as a rapid divergence of two groups after long periods of no change.

1972, when N. Eldredge and S. J. Gould suggested that speciation itself, and the morphological changes accompanying speciation, occur rapidly, separated by long periods of time when little change occurs (*stasis*). They called their model *punctuated equilibrium* (periods of stasis punctuated by rapid evolutionary change). Although in practice the models are very hard to tell apart. They both start with the same ancestral species and predict the same number of modern species. Allopatric, parapatric, and sympatric speciation mechanisms apply to both punctuated equilibrium and phyletic gradualism. The only major difference between the models is the rate of change, and this can only be discovered from an almost complete fossil record. The punctuated equilibrium model has brought much excitement to modern evolutionary biology. We await a time in the near future when we can decide which model has predominated in evolutionary history.

Genetic Variation

Darwinian evolution depends on the variation within a population. E. B. Ford, a British evolutionary biologist, applied the term *genetic polymorphism* to the occurrence of more than one allele at a given locus. Usually, we consider a locus polymorphic if a second allele occurs in the population at a frequency of 5% or more. Before the mid-1960s, the general belief was that only a few loci were polymorphic in any individual or any population.

In 1966, two researchers found a way to sample the genome in what they perceived was a random manner. R. C. Lewontin and J. L.

Hubby used acrylamide-gel electrophoresis to investigate variability in a fruit fly species, *Drosophila pseudoobscura*. (H. Harris reported independent, similar work with human DNA.) Lewontin and Hubby reasoned that choosing enzymes and general proteins that are amenable to separation by electrophoresis, is, in fact, choosing a random sample of the genome of the fruit fly. If this is the case, then the degree of polymorphism found by electrophoretic sampling would provide an estimate of the amount of variability occurring in the individual organism and in the population. Their results were startling.

Lewontin and Hubby found that the species was polymorphic at 39% of eighteen loci examined, the average population was polymorphic at 30% of its loci, and the average individual was heterozygous at 12% of its loci. The high rate of polymorphism sparked two interrelated controversies. The first was whether electrophoresis does, in fact, randomly sample the genome. The second was whether most electrophoretic alleles are maintained in the population by natural selection. Let us return to the arguments after looking at ways in which genetic polymorphisms could be maintained in natural populations.

Maintaining Polymorphisms

Heterozygote advantage

When selection acts against both homozygotes, an equilibrium is achieved, dependent solely on the selection coefficients, that maintains both alleles. The classic example of heterozygote advantage in human beings is sickle-cell anemia. Sickle-cell hemoglobin (Hb^s) differs from normal hemoglobin (Hb^A) because it has a valine in place of a glutamic add in position number 6 of the beta chain of the globin molecule. When the availability of oxygen is reduced, the erythrocytes containing sickle-cell hemoglobin change from round to sickle-shaped cells. There are two unfortunate consequences: (1) sickle-shaped cells are rapidly broken down, which causes anemia as well as hypertrophy of the bone marrow, and (2) the sickle cells clump, which blocks capillaries and produces local losses of blood flow that result in tissue damage.

This condition of reduced fitness would lead one to predict that the sickle-cell allele would be selected against in all populations and, therefore, would be rare. But this is not the case. The sickle-cell allele is common in many parts of Africa, India, and southern Asia. What could possibly maintain this detrimental allele? In the search for an answer to this question, biologists discovered that the distribution of the sickle-cell allele coincided well with the distribution of malaria. The following facts have now been uncovered. The sickle-cell homozygote

(Hb^sHb^s) almost always dies of anemia. The sickle-cell heterozygote (Hb^AHb^s) is only slightly anemic and has resistance to malaria. The normal homozygote (Hb^AHb^A) is not anemic and has no resistance to malaria. Thus, in areas where malaria is common, the most fit genotype of the three appears to be the sickle-cell heterozygote, which has resistance to malaria and only a minor anemia.

This conclusion is supported by the changes in allelic frequencies that occur when a population from a malarial area moves to a nonmalarial area. Since the normal homozygote is no longer at risk for malaria, selection acts mainly on the sickle-cell homozygote and, to a slight extent, on the heterozygote. The African population is, of course, under malarial risk, whereas the American population is not. The sickle-cell hemoglobin allele (Hb^s) is reduced in frequency in African Americans.

Heterozygote advantage is an expensive mechanism for maintaining a polymorphism. Losses must occur in both homozygous groups in order for the polymorphism to exist. Thus, part of the reproductive output of a population is lost each generation to maintain each polymorphism under heterozygote advantage. In the case of sickle-cell anemia, this means a tragic loss of human life due to either anemia or malaria. (The loss of individuals to maintain genetic variation at a particular locus is called *genetic load*. In the sickle-cell case, it is due to the segregation of individuals with lowered fitness and is therefore called *segregational load*.) Very few other examples of heterozygote advantage have been documented.

Frequency-dependent selection

All the selection models discussed so far have had selection coefficients that were constants. This is not always the case. For example, L. Ehrman has shown that when a female fruit fly has a choice between mates with different genotypes, the female fly chooses to mate with a male with a rare genotype. *Frequency-dependent selection* is selection in which the fitnesses of genotypes change according to their frequencies in the population.

The population geneticist Bruce Wallace has coined the terms *hard selection* and *soft selection* to deal with cases of frequency and density dependence. (Density-dependent selection exists when the fitness of a genotype changes as population density changes. We will not deal with that here.) Wallace defined soft selection as selection in which the selection coefficients depend on the frequency and density of genotypes. Hard selection is selection that is independent of both frequency and density. For example, the low fitness of sickle-cell anemia homozygotes involves hard selection because of the objectively deleterious effects of the anemia. Soft selection could be envisioned as selection that might act on aggressive behavioural genotypes in some lemming and field mouse species. When population density and frequency of the genotypes are low, these animals survive and reproduce. As population density increases, there can be a selection for more aggressive genotypes because they may be more successful in obtaining resources. As density increases further and the frequencies of the aggressive genotypes increase, they may be selected against because of the preoccupation of these aggressive individuals with territory defense under crowded conditions. This has been suggested as a mechanism of wildlife's "lemming cycle," rapid declines in the density of lemming and field mouse populations every three to five years.

A model for frequency-dependent selection can be constructed by assigning fitnesses that are not constants. One way to do this is to assign fitnesses that are a function of allelic frequencies. Thus, the assigned fitnesses for one locus with two alleles could be $(1.5 - p)$, 1, and $(1.5 - q)$ for the *AA*, *Aa,* and *aa* genotypes, respectively. An interesting outcome of this model is that at $p = q = 0.5$, the system is in equilibrium, and no selection takes place because all the fitnesses are equal to 1.

Another way of looking at frequency-dependent selection is to look at the situation in which each genotype exploits a slightly different resource. As a genotype becomes rare, competition for the resource that genotype uses will likely decrease, and the genotype will thus have an advantage over the common genotypes, which are competing for resources. This type of selection is probably very common.

Transient polymorphism

A genetic polymorphism can result when an allele is being eliminated either by random or selective mechanisms. If a population starts out homozygous for the *a allele,* for example, and a mutation brings in a more favoured *A* allele, the population gradually becomes all *A* through directional selection. However, during the process of replacement, both alleles are present.

Other systems

Selection at one stage in the life cycle of an organism can balance a different form of selection at another stage in the life cycle. For example, an allele can be favoured in a larva but selected against in an adult. There can also be a balance of selection in different parts of the habitat in a heterogeneous environment. For instance, an allele can be favoured in a wet part of the habitat but selected against in a dry part.

Maintaining many Polymorphisms

In summary, allelic polymorphisms in a population were classically accounted for by heterozygote advantage, frequency-dependent selection, or, infrequently, some other mechanism. Until Lewontin and Hubby did their work, heterozygote advantage was considered the most common method of maintaining a polymorphism at a given locus. The maintenance of an allele by heterozygote advantage costs the population a certain number of its offspring due to the mortality (or sterility) of the homozygotes. Most populations can afford the loss if polymorphisms are maintained at only a few loci. After Lewontin and Hubby reported that polymorphisms seemed to exist at a large proportion of loci, new explanations were needed to account for them. Three explanations were considered:

1. Electrophoresis (the technique used in Lewontin and Hubby's research) does not randomly sample the genome, and thus the large amount of variability they found does not really exist.
2. New population genetic models can be derived that explain how natural selection maintains this large amount of variability
3. Electrophoretic alleles are not under selective pressure. That is, allozymic forms of an enzyme all perform the function of the enzyme equally well. This idea is called the *neutral gene hypothesis*.

Sampling the genome

Does electrophoresis randomly sample the genome? Since, on the basis of DNA content, the genome of higher organisms has the potential to contain half a million genes, this question may be difficult to resolve. Since the original reports of Lewontin and Hubby and Harris, numerous studies on many different organisms agree, for the most part, on the high amount of polymorphism in natural populations. However, several lines of evidence suggest that the results from electrophoresis are actually underestimates of the true amount of genetic variability present in a population.

The majority of amino acid substitutions, for example, do not change the charge of the protein. Thus, what appear to be single bands on an electrophoretic gel could actually be heterogeneous mixtures of the products of several alleles. Also, we now know that glycolytic enzymes are less polymorphic than other enzymes. Since glycolysis is a limited process in which most enzymes are not involved, it follows that the average heterozygosity over all loci should be slightly higher than the original estimates that included glycolytic enzymes. Recent technical advances of multidimensional electrophoresis and DNA sequencing support the hypothesis that electrophoresis does randomly sample the genome. However, DNA sequencing studies have shown that abundant variation exists, especially in the third (wobble) position of codons, and in parts of introns. Heterozygosity at the DNA sequence level seems to approach 100%.

Multilocus selection models

Can standard genetic models account for the high degree of variability in natural populations? If each locus is considered independently, then for each polymorphic locus, offspring in a population lost to maintain that polymorphism by heterozygote advantage are independent of offspring lost due to selection at other loci. The losses would soon outstrip the reproductive capacity of any species. Models proposed since Lewontin and Hubby's report have suggested that natural selection favours the individuals that are the most heterozygous overall. Individuals selected against because of their homozygosity would be individuals with many homozygous loci. In other words, natural selection acts on the entire genome, not on each locus separately. We can show algebraically that the large number of polymorphisms that exist in natural populations could be maintained according to these models.

Neutral alleles

The high incidence of polymorphism that electrophoresis reveals may not be important from an evolutionary point of view. If all or most electrophoretic alleles are neutral (i.e., if no allele is more fit than its alternative) or only very slightly deleterious, there is virtually no selection at these loci, and the variation observed in the population is merely a chance accumulation of a combination of mutation and genetic drift. This model, proposed by M. Kimura of Japan, is an alternative to the natural selection model.

Which Hypothesis is Correct?

Researchers who favour the concept that most electrophoretic alleles are neutral do not deny that selection exists. They do not hold that evolution is non-adaptive, but say merely that most of the molecular variation (electrophoretic) found in nature is not related to fitness—it is neutral. Thus, the demonstration that selection actually exists, in electrophoretic systems or otherwise, is not proof against the neutralist view. No one denies the explanation for the maintenance of sickle-cell anemia. Selection at several other electrophoretic systems is also known.

For example, R. Koehn showed that different alleles of an esterase locus in a freshwater fish in Colorado produced proteins with different enzyme activities at different water temperatures. Koehn then showed that the alleles were distributed as one would predict on the basis of the water temperature. In other words, the distribution of alleles correlated with the distribution of water temperature. The enzyme produced by the *ES*-1[a] allele functioned best at warm temperatures, whereas the enzyme produced by the *ES*-1[b] allele functioned best at cold temperatures. The cold-adapted enzyme was prevalent in the fish in colder waters (higher latitudes), and the warm-adapted enzyme was prevalent in the fish in warmer waters (lower latitudes).

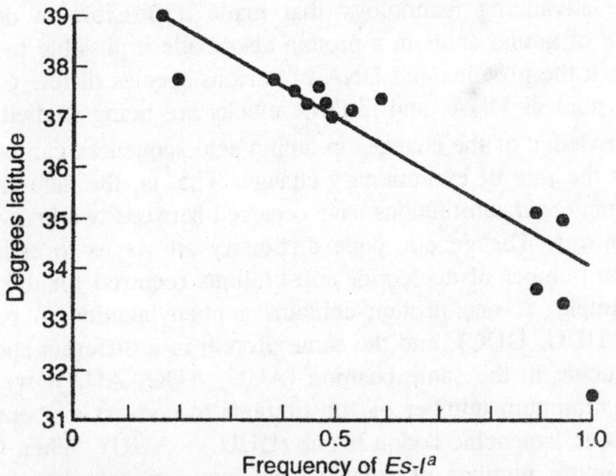

Fig. 2.5. Relation of latitude and frequency of the warm-adapted esterase allele Es-I[B] *in populations of the fish Catostomus clarki. Note how the frequency of the allele increases as latitude decreases (warmer water).*

Isolated instances of selection, however, do not adequately prove the case for maintaining variation by means of natural selection or disprove the case for maintaining variation of neutral alleles. Both theories recognize natural selection as the guiding force in producing adapted organisms. What is needed is proof that the majority of polymorphic loci are either being selected or are neutral. For this proof, many loci must be examined independently—a very difficult undertaking—or some grand pattern must emerge supporting one hypothesis or the other.

Grand Patterns of Variation

Clinal selection

Data on the geographic distribution of alleles fail to adequately support either theory. Often, a single allele predominates over the range of a species. Changes in allelic frequency from one geographic area to another can often be attributed to *clinal selection*, selection along a geographic gradient, in which allelic frequencies change as altitude, latitude, or some other geographic attribute changes. The general increase in the *Es*-5^b frequency from west to east in the southern United States. But, in line with the neutralist view, geographic patterns can also be produced by neutral alleles with a very low level of migration, as little as one individual per one thousand per generation.

Molecular evolutionary clock

The advancing technology that made it possible to detect the sequence of amino acids in a protein also made it possible to discover how much the proteins and DNA of various species differ. Currently, protein, nuclear DNA, and mtDNA clocks are being studied.

Knowledge of the changes in amino acid sequences can be used to estimate the rate of evolutionary change. That is, the data show how many amino acid substitutions have occurred between two known groups of organisms. The genetic code dictionary allows us to estimate the minimum number of nucleotide substitutions required for this change. For example, if one protein contains a phenylalanine in position 7 (codons UUU, UUC), and the same protein in a different species has an isoleucine in the same position (AUU, AUC, AUA), we can see that the minimum number of substitutions to convert a phenylalanine codon to an isoleucine codon is one (UUU → AUU). When we know the minimum number of substitutions, we can calculate molecular *evolutionary rates*, nucleotide substitutions per million years. In a sense, these rates provide us with a *molecular evolutionary clock* that measures evolutionary time in nucleotide substitutions.

Many studies of the rate of amino add and nucleotide substitutions have been done on hemoglobin, on cytochrome *c,* on a class of proteins involved in blood clotting called fibrinopeptides, and on many others. From comparisons of this type, we can calculate the actual number of amino add differences, as well as percentage differences. This type of information can be used two ways.

First, we can construct a *phylogenetic tree* that tells us the evolutionary history of the species under consideration. This tree can be compared with phylogenetic trees constructed by more classical

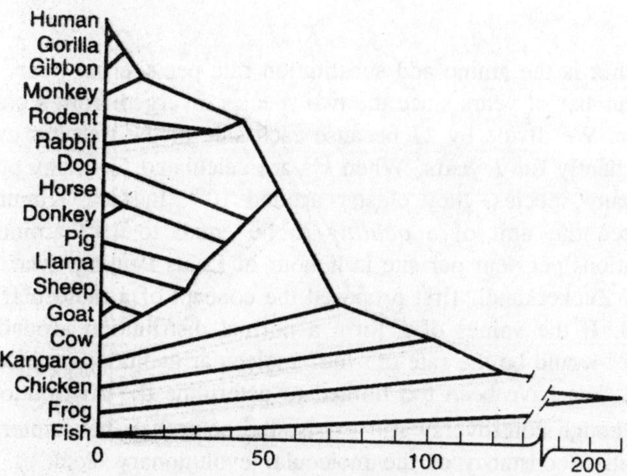

Fig. 2.6. Composite evolution of hemoglobin, cytochrome c, and fibrinopeptide A. The total number of nucleotide substitutions appears on the horizontal axis.

means using fossil evidence and evidence from morphology, physiology, and development. From the comparisons, we can look at areas of disagreement in an attempt to find out the best way to create phylogenetic trees. In addition, molecular phylogenies can give us information unattainable in any other way, as, for example, when the fossil record is incomplete or ambiguous.

A second use of DNA or amino add difference data is to determine average rates of substitution. Once we know the current amino acid differences in the proteins of two species, it is possible to estimate the actual number of nucleotide substitutions that have taken place over evolutionary time using the statistical Poisson distribution, which deals with rare events. The index, *K,* is the average number of amino acid substitutions, per site, between two proteins:

$$K = -\ln(1 - p)$$

in which ln is the natural logarithm (to the base *e*), and $p = d/n$ in which d is the number of amino acid differences and n is the total number of amino acid sites being compared. For example, $n = 8$ and $d = 3$ between the dog and chicken. Thus

$$K = -\ln(1 - 0.375) = 0.47$$

Therefore, the average number of amino acid substitutions, per site, between dog and chicken is 0.47.

We can take this calculation one step further by determining the per-year rate:

$$k = K/2T$$

in which k is the amino add substitution rate per site per year, and T is the number of years since the two species diverged from a common ancestor. We divide by $2T$ because each side of the tree has evolved independently for T years. When k's are calculated for many proteins over many species, they cluster around 10^{-9}. In fact, Kimura has suggested the unit of a *pauling* to be equal to 10^{-9} amino acid substitutions per year per site in honour of Linus Pauling, who, along with E. Zuckerkandl, first proposed the concept of a molecular clock in 1963. If the values of k form a normal distribution around 10^{-9}, then 10^{-9} would be the rate of "the" molecular evolutionary clock. So far, the data have been too limited to determine the distribution.

Although controversy still exists, the neutralists have interpreted the relative constancy of the molecular evolutionary clock as strong evidence in support of the neutral gene hypothesis. A constant rate of molecular evolution over many groups of organisms over many different time intervals implies that the substitution rate is a stochastic or random process rather than a directed or selectional process. This is not to say that no adapted changes occur in proteins or that there are no constraints. In fact, the evidence suggests that three classes of amino acids can be grouped in terms of substitution rate: invariant, moderately variant, and hypervariant. It seems possible that virtually no substitutions of amino acids will occur in and around the active site of the enzyme since any amino acid change in that area might be deleterious or lethal. For example, a segment of cytochrome *c* that runs from amino adds 70 to 80 is invariant in all organisms tested. This area includes a binding site of the protein.

DNA variation

If the neutralist view of molecular evolution is correct, we should be able to make some predictions about rates of change in DNA. For example, we predict that DNA under greater constraint should amass fewer base changes than DNA under lesser constraint. We could test this by looking at the accumulation of mutations in the three positions of the codon, or we could look at DNA that is not directly translated, such as pseudogenes or introns, which are probably under lesser constraint. Let us first look at the three positions of the codon.

A reexamination of the codon dictionary shows that the third, or wobble, position of the codon should be under less constraint. Eight amino adds belong to unmixed families; their amino adds are defined by the first and second positions coupled with any of the four bases in the third position of the codon. The remaining amino acids belong to mixed families; the first two positions and the purine or pyrimidine nature of the third position in their codons is important. Hence, the wobble (third) position of the codon is under the least constraint and should build up the most neutral or near-neutral mutations.

In addition, analysis of changes in the first and second positions indicates that more drastic change takes place by mutation of the second rather than the first position of the codon. Thus, we predict that evolutionary distance, as measured by base substitutions, should be greatest for the third codon position and least for the second position. This turns out to be generally true.

It should be clear that a major problem facing those who study evolutionary clocks is how to calibrate them. Are average changes uniform throughout lineages? Do clocks speed up, slow down, or show other unpredictable changes through time? There is evidence, for example, that both the nuclear and mitochondrial DNA clocks have slowed down in the hominid lineage as compared with old world monkeys. If the clocks change speed in different lineages, at different times, and for different parts of the genome, there will be errors in interpreting lineages and errors in using averages to understand the general patterns of change.

At this point, it is probably safe to say that while natural selection acts to create organisms that are adapted to their environments, many nucleotide and amino acid changes may not have measurable effects on the fitness of the organism, and hence their frequencies may be determined by the stochastic processes of mutation and genetic drift. Adaptation is by natural selection, but neutral variation most certainly also occurs in organisms.

Sociobiology

We close this chapter by looking at a level of evolution only recently addressed. In 1975, E. O. Wilson published a mammoth tome entitled *Sociobiology*: *The New Synthesis*. This book has been at the center of major controversies that have spread to the fields of sociology, psychology, anthropology, ethology, and political science.

Altruism

V.C. Wynne-Edwards published a book in 1962 entitled *Animal Dispersion in Relation to Social Behaviour*. In it, he suggested that animals regulate their own population density through altruistic behaviour. For example, under crowded conditions, many birds cease reproducing. The interpretation of this phenomenon was that these birds were being altruistic: Their failure to breed was for the ultimate good of the species. (*Altruism* means risking loss of fitness in an act that could improve the fitness of another individual.) Wynne-Edwards suggested a mechanism called *group selection*: groups that had altruistic behaviour would have a survival advantage over groups that did not.

In 1966, G. Williams, in his book *Adaptation and Natural Selection: A Critique of Some Current Evolutionary Thought,* refuted the altruistic view with the charge that individuals that performed altruistic acts would be selected against. In other words, organisms not performing altruistic acts would have a higher degree of fitness. Williams held that apparent altruism had to be interpreted on the basis of benefits accruing to the individual performing the altruistic act. After his book, the idea of doing something for the good of the species became passe. How, then, can apparent altruism be accounted for? How can we explain why ground squirrels appear to put themselves at risk to predators by giving alarm calls, and why female workers in ant, wasp, and bee colonies forsake reproduction in order to work for the colony? *Sociobiology*, the study of the evolution of social behaviour, attempts to answer these questions.

Kin Selection and Inclusive Fitness

In 1964, W.D. Hamilton developed concepts that explained altruistic acts without resorting to group selection. Starting with the known fact that relatives have alleles in common, Hamilton suggested that natural selection would favour an allele that promoted altruistic behaviour toward relatives because the result might be an increase in copies of that allele in the next generation. The proportion of alleles shared by two individuals can be defined as a *coefficient of relationship*, r. If an individual has a certain allele, the probability that a particular relative also has that allele is r. Siblings have an $r = 1/2$. A squirrel is likely to have virtually all its alleles still viable if it sacrifices itself for two or more siblings. In fact, natural selection should definitely favour altruism of an individual toward three siblings because, in a sense, natural selection is weighing 1 copy of an individual's alleles (the individual itself) versus 1.5 copies (three siblings).

This sort of reasoning has been termed the *calculus of the genes*. It does not imply that individuals actually think these things out; rather, natural selection has favoured the individuals that behave this way. Hamilton referred to the sum of an individual's fitness plus the fitness effects of alleles that relatives share as *inclusive fitness*. He referred to the way natural selection acts on inclusive fitness as *kin selection*.

Hamilton applied his ideas of inclusive fitness and kin selection to explain sterile castes in the eusocial (truly social) hymenoptera (bees, ants, and wasps). The workers in these colonies are sterile females. Why do they forsake their ability to reproduce in order to help maintain the hive or colony? The answer seems to come from *haplodiploidy*, the unusual sex-determining mechanism of these species. In the eusocial hymenoptera with sterile castes, fertilized eggs produce diploid females, whereas unfertilized eggs produce haploid males (drones). The *difference* between a reproductive queen and a sterile worker in bees is larval nutrition: larvae fed "royal jelly" can become queens. Hamilton showed that since a worker is more closely related to her sisters than to her own potential offspring, kin selection could favour a worker who helps her sisters at the expense of her own reproduction.

Figure shows a queen (female) with alleles A_1 and A_2 at the A locus and a haploid drone (male) with the A_3 allele. A daughter will have either the A_1A_3 or A_2A_3 genotype. If we compare one of these daughters with her sisters, we see that the average r = 0.75—half of the time, r = 1.0, and the other half of the time, r = 0.5. A queen and her daughters have an r = 0.5. Thus, we see that workers (females)

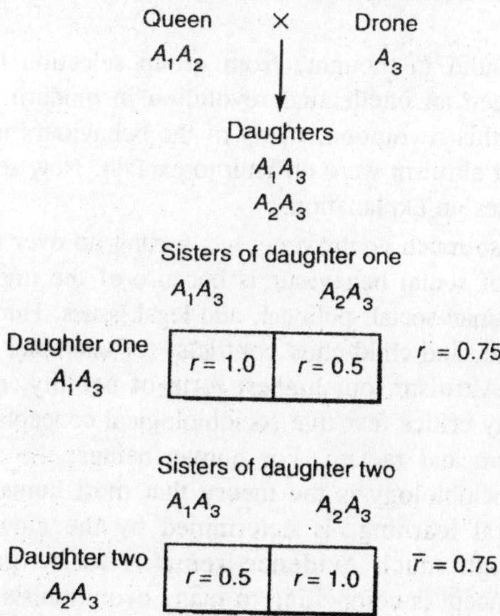

Fig. 2.7. Haplodiploidy in eusocial hymenoptera produces sisters with an average r of 0.75. Because drones (males) are haploid, queens produce daughters of only two genotypes at any locus.

are more closely related to their sisters, and hence are at a reproductive advantage by raising them rather than their own young. Wilson has pointed out that sterile caste systems have evolved among insects in only one other group beside the eusocial hymenoptera, the termites. Although eusocial hymenoptera make up only 6% of insects, sterile castes have independently evolved at least eleven times. This is compelling evidence for the validity of Hamilton's analysis. Only one noninsect example of a caste has been discovered: the naked mole rat, a small subterranean rodent living in Africa, has this type of social system.

Many studies concerned with apparently altruistic acts have provided a large body of support for Hamilton's theory of kin selection and inclusive fitness. P. Sherman, working with ground squirrels, for example, has observed that the individuals that make the alarm calls have the most to gain from the standpoint of inclusive fitness; these individuals are resident females surrounded by kin.

One other explanation for altruism is also consistent with benefits to individual fitness. It is that many apparently altruistic acts are in reality selfish—they just look altruistic. To be altruistic, an individual must risk reducing its fitness to potentially benefit the fitness of others. We may, in fact, misinterpret some acts as altruistic that simply are not.

This turnaround in thought, from group selection to individual selection, has been an intellectual revolution in modern evolutionary biology. Before this revolution, many of the behaviours in nature that involved apparent altruism were difficult to explain. Now sociobiological reasoning provides an explanation.

The reason so much controversy has sprung up over the theory of genetic control of social behaviour is because of the implications the theory has for human social, political, and legal issues. Human husband-wife, parent-child, and child-child conflicts, for example, may be built into the genes. Altruism, our highest form of nobility, may be mere selfishness. Many critics fear that sociobiological concepts can be used to support sexism and racism. For human beings, the alternative to the theory of sociobiology is the theory that most human behaviour, including cultural learning, is determined by the environment. At present, although much evidence remains to be gathered, the sociobiology concept is compelling to many evolutionists.

3

Evolutionary Advantage of Sex

In this chapter, we use our knowledge of population and quantitative genetics to explore one the most compelling of evolutionary conundrums: Why sex? Rather than attacking all aspects of this question, we restrict our attention to sex in eukaryotic anisogamous creatures, species where the female produces relatively large gametes, such as eggs or seeds, and the male produces a plethora of small and mobile gametes, such as sperm or pollen. This describes, in other words, all of the animals and higher plants, most of which indulge in sexual reproduction. However, there are species that do not. Most could be called parthenogenetic. There are many forms of parthenogenesis. In forms where meiosis is not involved, like apomixis and endomitosis, offspring are genetically almost identical to their parent. The differences that do occur are caused by mutation or mitotic recombination. In forms of parthenogenesis where meiosis is involved, like automixis, offspring are not identical to their parent and are usually inbred. Here we will be concerned only with the first form of parthenogenesis, which may be found in many major groups of plants and animals. However, except for one group of rotifers, there are no major taxonomic groups that are entirely parthenogenetic. Parthenogenetic species must inevitably face quick extinction due, presumably, to problems associated with the absence of sex, despite the fact that parthenogenetic species enjoy a twofold advantage over their sexual siblings.

Each female, whether sexual or not, will leave behind, on average, two offspring. On average, in sexual species, one of the two offspring

Fig. 3.1. A comparison of the consequences of sexual versus parthenogenetic reproduction.

will be a female and the other a male. Each of these offspring carries one haploid complement of the female's genetic material; the other haploid genome comes from the male parent. A parthenogenetic female will also leave behind two offspring, but both will be females because they are clones of their mother. Moreover, all of the genetic material, four haploid complements, comes from the mother. Thus, the parthenogenetic female leaves behind twice as much of her genetic material as does the sexual female.

Imagine a sexual species in which a dominant mutation arises that causes parthenogenesis. The genome carrying the mutation will appear in two individuals in the next generation, four the generation after that, and so on, doubling in number each generation. Meanwhile, all genomes without the mutation will be replacing themselves once each generation, on average. Eventually, the species will be made up entirely of parthenogenetic individuals. This is the twofold cost of sexual reproduction.

There must be some problem with this scenario or else all species would become parthenogenetic. One possibility, of course, is that mutations to parthenogenesis either do not occur or are so deleterious that they swamp their twofold advantage. However, in many groups parthenogenesis does recur, so we must look elsewhere for an explanation of its ultimate failure.

Sex in diploids is associated with segregation and meiotic crossing-over; parthenogenetic species have neither. Thus, a natural place to look for the advantages of sex is the examination of segregation and recombination. Both, as we shall see, can confer certain advantages to a species.

Genetic Segregation

Although the reasons for sex have been discussed for decades, most of the arguments centered on crossing-over and recombination.

Only recently did Mark Kirkpatrick and Cheryl Jenkins point out that segregation itself can confer an advantage. Their 1989 paper, "Genetic segregation and the maintenance of sexual reproduction," points out that the substitution of an advantageous mutation in a parthenogenetic species requires two mutations in a lineage, whereas substitution in a sexual species requires only one mutation. If the evolutionary success of a species depends on its ability to evolve rapidly and if advantageous mutations are relatively rare, then it is easy to imagine that the parthenogens will be at a disadvantage.

While plausible, this argument carries little force unless it can be shown to enhance the lot of sexual species sufficiently to overcome their twofold cost of sex. Kirkpatrick and Jenkins did this by first finding the number of loci in a parthenogenetic species that are heterozygotes waiting for a second mutation that allows the fixation of the advantageous mutation and then calculating the cost of being heterozygous at these loci.

With incomplete dominance, the fixation of an advantageous mutation in a parthenogen occurs in two steps. In the beginning, a typical locus is imagined to be fixed for the A_2 allele with the fitnesses of the genotypes as follows:

Genotype:	A_1A_1	A_1A_2	A_2A_2
Fitness:	$1 + s$	$1 + hs$	1

The A_1 allele is favoured, so $0 < s$. The rate of mutation to the A_1 allele is u. Each generation, $2Nu$ A_1 mutations enter the population, on average, and a fraction $2sh$ of these escape loss by genetic drift. Thus, the rate of entry of A_1 alleles into the population is $4Nuhs$. If there are L homozygous loci experiencing this sort of selection, then the rate of conversion from homozygotes to heterozygotes is

$$4NuLhs. \qquad \ldots(1)$$

The environment is assumed to be changing in such a way that L is constant rather than decreasing with the substitution of each advantageous mutation.

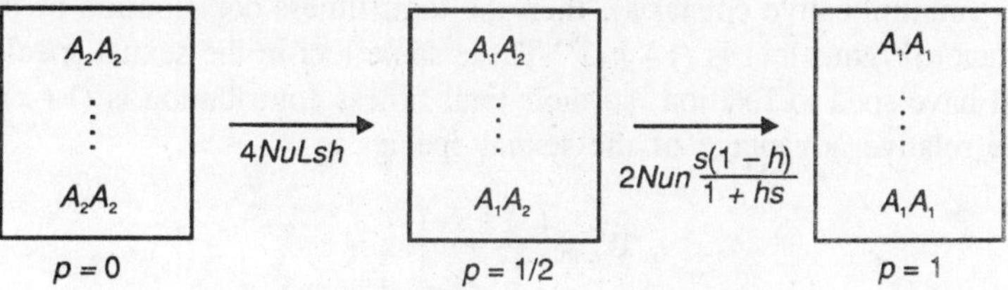

Fig. 3.2. Evolution in parthenogens.

The second stage is the conversion of A_1A_2 heterozygotes to A_1A_1 homozygotes, which requires the fixation of an A_1 allele on the chromosomes with the A_2 allele. The mean number of A_1 mutations entering each generation at one locus is *Nu*. (The factor of two is missing because each individual is an A_1A_2 heterozygote.) The survival probability of these mutations is twice the selective advantage of A_1A_1 over A_1A_2. The selection coefficient of an A_1A_1 homozygote relative to the A_1A_2 heterozygote, call it s', is found by solving

$$1+e' = \frac{1+e}{1+hs}$$

to obtain $s' = s(1 - h)/(1 + hs)$. Thus, the rate of conversion of a particular locus from a population of heterozygotes to one of homozygotes is

$$2Nus(1 - \text{h})/(1 + hs).$$

If n loci are currently heterozygous, then the rate of conversion from A_1A_2 to A_1A_1 for the genome is

$$2Nun\frac{s(1-h)}{1+hs}. \qquad ...(2)$$

The equilibrium number of heterozygous loci is found by setting the rate of conversion to heterozygotes, Equation 1, equal to the rate of conversion away from heterozygotes, Equation 2, and solving for n,

$$\hat{n} = \frac{2Lh(1+hs)}{1-h},$$

which is Equation 1 in the Kirkpatrick and Jenkins paper.

In sexual species, the A_1 allele sweeps through the population without waiting for a second mutation. The asexual species is at a relative disadvantage while sitting on the plateau waiting for the second mutation to sweep through.

The fitness contribution of each of then heterozygous loci to the parthenogenetic individual's overall fitness is $1 + hs$. If the overall fitness of an individual is the product of the fitnesses of individual loci (multiplicative epistasis), then the total fitness contribution of the $\hat{n}$ heterozygous loci is $(1+hs)^{\hat{n}}$. These same loci in the sexual species will have sped to fixation, so their total fitness contribution is $(1+s)^{\hat{n}}$. The relative advantage of the sexual species is

$$W_s = \left(\frac{1+s}{1+hs}\right)^{\hat{n}},$$

which is Equation 2 in the Kirkpatrick and Jenkins paper.

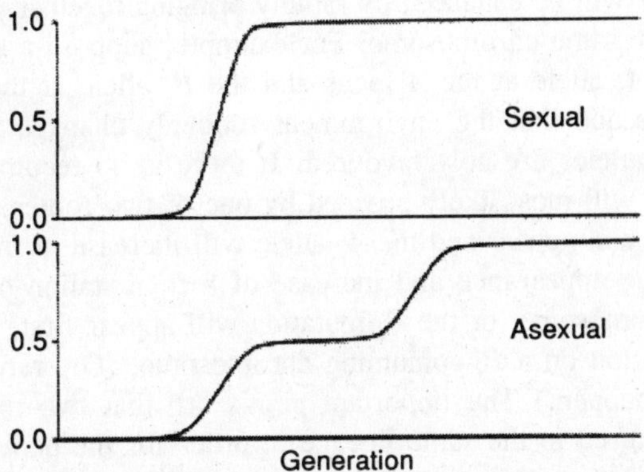

Fig. 3.3. The trajectory of allele frequencies under directional selection in asexual and sexual species.

Suppose, for example, that $s = 0.01$, $L = 100$, $h = 1/2$; then the number of heterozygous loci is $\hat{n} = 201$ and the relative advantage of the sexual species is $W_s = 2.7$, which is sufficient to overcome the twofold cost of sex.

If a species is constantly evolving, then sex with its accompanying segregation will speed up the fixation of advantageous mutations and thereby raise the mean fitness of the population. We have no idea about plausible values of L and s for natural populations, so we cannot say with any confidence that sex is maintained due to the benefits of segregation. We can say with confidence that segregation could explain the maintenance of sex.

There are some reservations about Kirkpatrick and Jenkins' model, though not about the basic idea. Foremost among these is the assumption that the total fitness of an individual is obtained by multiplying the fitness contributions of individual loci. The experimental evidence, as will be shown later, suggests that fitness drops off faster than this as loci interact, the quantitative effects of which will be complex, affecting both $\hat{n}$ and W_s but not affecting the fundamental conclusion that segregation speeds up the rate of substitution of incompletely dominant mutations. Other reservations may be found in the paper itself.

Crossing-over

In addition to segregation, sex is also associated with crossing-over and its consequence, recombination. As an evolutionary force, recombination causes alleles at different loci eventually to become randomly associated with one another. There is an obvious evolutionary advantage to recombination if there are frequent situations where a species' fitness will be enhanced by rapidly bringing together particular alleles onto the same chromosome. For example, suppose a species is fixed for the A_2 allele at the A locus and the B_2 allele at the closely linked B locus and that the environment suddenly changes such that the A_1 and B_1 alleles are now favoured. If there is no recombination, then evolution will most likely proceed by one of two routes. Either a mutation to A_1 will appear and the A_1 allele will increase in frequency, followed by the appearance and increase of a B_1 mutation on an A_1-containing chromosome, or the B_1 mutation will appear first, followed by an A_1 mutation on a B_1-containing chromosome. (Or, rarely, both events could happen.) The important points are that two mutational events are required in the same lineage in order for the population to achieve fixation at both loci and that the second mutational event is unlikely to happen until the frequency of the chromosomes carrying the first mutation is very high.

If recombination is allowed and if both the A_1 and B_1 mutations appear and increase, recombination can cause the two to come together on the same chromosome without requiring a second mutation on the same lineage. Moreover, this can occur before either A_1 or B_1 mutations are very frequent. Like segregation, recombination appears to speed up evolution. In the case of recombination, the speedup is due to the production of a recombinant gamete that is more fit than other gametes in the population.

To appreciate some of the subtleties of this argument, which was published in 1932 by the geneticist H. J. Muller, we need to learn more about the evolutionary consequences of recombination. The simplest model capable of showing the effects of recombination is of a diploid species with two linked loci, each with two segregating alleles. The probability that a recombinant gamete is produced at meiosis is denoted by r, which is often called the recombination rate. (The genetic or map distance between the loci is always greater than r because it is the average number of recombinational events rather than the probability of producing a recombinant offspring.)

There are four gametes in the population, A_1B_1, A_1B_2, A_2B_1, and A_2B_2 with frequencies p_{11}, p_{12}, p_{21}, and p_{22}, respectively. The frequency of the A_1 allele, as a function of the gamete frequencies, is $p_{1\cdot} = p_{11} + p_{12}$. The dot in the subscript acts as a placeholder to remind you that there is another locus around, but that we are ignoring it. Similarly, the allele frequency of B_1 is $p_{\cdot 1} = p_{11} + p_{21}$. Recombination changes the frequencies of these gametes in a very simple way. For example, the frequency of the A_1B_1 gamete after a round of random mating, p'_{11}, is simply

$$p'_{11} = (1-r)p_{11} + rp_{1\cdot}p_{\cdot 1}. \qquad ...(3)$$

This expression is best understood as a statement about the probability of choosing an A_1B_1 gamete from the population. A randomly chosen gamete will have had one of two possible histories: Either it will be a recombinant gamete (this occurs with probability r) or it won't be (this occurs with probability $1 - r$). If it is not a recombinant, then the probability that it is an A_1B_1 gamete is p_{11}. Thus, the probability that the chosen gamete is an unrecombined A_1B_1 gamete is $(1 - r)p_{11}$, which is the first term on the right side of Equation 3. If the gamete is a recombinant, then the probability that it is A_1B_1 is the probability that the A locus is A_1, which is just the frequency of A_1, $p_{1\cdot}$, times the probability that the B locus is B_1, $p_{\cdot 1}$. The probability of being a recombinant gamete and being A_1B_1 is $rp_{1}p_{\cdot 1}$, which is the rightmost term of Equation 3. The allele frequencies can be multiplied because the effect of a recombination is to choose the allele at the A and B loci independently.

The change in the frequency of the A_1B_1 gamete in a single generation of random mating is, from Equation 3,

$$\Delta_r p_{11} = r(p_{1\cdot}p_{\cdot 1} - p_{11}). \qquad ...(4)$$

The equilibrium gamete frequency is obtained by solving $\Delta_r p_{11} = 0$,

$$\hat{p}_{11} = p_{1\cdot}p_{\cdot 1}.$$

If there were no tendency for the A_1 allele to be associated with the B_1 allele, the probability of choosing an A_1B_1 allele from the population would be the product of the frequencies of the A_1 and B_1 alleles. As this is precisely the equilibrium state of the population, we conclude that recombination removes associations between alleles on chromosomes. The rate of change of the gamete frequency due to recombination is simply the recombination rate, r.

The deviation of the frequency of A_1B_1 from its equilibrium value is called *linkage* disequilibrium,

$$D = p_{11} - p_{1\cdot}p_{\cdot 1}. \qquad ...(5)$$

Thus, the frequency of A_1B_1 gamete may be written

$$p_{11} = p_{1\cdot}p_{\cdot 1} + D,$$

which emphasizes that the departure of the gamete frequency from its equilibrium value is determined by D. Obviously, at equilibrium $D = 0$. The linkage disequilibrium may also be written in the more conventional form

$$D = p_{11}p_{22} - p_{12}p_{21},$$

which leads to the following new expressions for the gamete frequencies:

Gamete:	A_1B_1	A_1B_2	A_2B_1	A_2B_2
Frequency:	p_{11}	p_{12}	p_{21}	p_{22}
Frequency:	$p_1p_{\cdot 1} + D$	$p_1p_{\cdot 2} - D$	$p_2p_{\cdot 2} - D$	$p_2p_{\cdot 2} + D$

The A_1B_1 and A_2B_2 gametes are often called *coupling gametes* because the same subscript is used for both alleles. The A_1B_2 and A_2B_1 gametes are called *repulsion gametes*. Linkage disequilibrium may be thought of as a measure of the excess of coupling over repulsion gametes. When D is positive, there are more coupling gametes than expected at equilibrium; when negative, there are more repulsion gametes than expected.

The value of D after a round of random mating may be obtained directly from Equation 3 by using $p_{11} = p_1p_{\cdot 1} + D$,

$$p_1'p_{\cdot 1}' + D' = (1-r)(p_1 p_{\cdot 1} + D) + rp_1 p_{\cdot 1}.$$

A few quick cancellations yield

$$D' = (1-r)D.$$

Some of the cancellations use the Hardy-Weinberg truism that allele frequencies don't change with random mating. We would be in trouble if the addition of loci affected the Hardy-Weinberg law for single loci!

The change in D in a single generation is

$$\Delta_r D = -rD,$$

which depends on the gamete frequencies only through their contributions to D. Finally,

$$D_t = (1-r)^t D_0,$$

showing, once again, that the ultimate state of the population is $D = 0$. Note that with free recombination ($r = 1/2$) the linkage disequilibrium does not disappear in a single generation. If you find this startling, follow D for a couple of generations in a population initiated with $p_{11} = p_{22} = 1/2$ and $r = 1/2$.

In natural populations, the reduction in the magnitude of linkage disequilibrium by recombination is opposed by several evolutionary forces that may increase $|D|$. Natural selection will increase D if selection favours coupling gametes over repulsion gametes or decrease D if repulsion gametes are favoured. Migration may increase the absolute value of D if the allele frequencies of the immigrants differ from those of the resident population. Finally, genetic drift can lead to changes in D due to random sampling. As recombination is most effective for loosely linked loci, we would expect to find tightly linked loci farther from linkage equilibrium than loosely linked loci. This has been seen in molecular data, but the linked loci must be very close indeed in order to see significant levels of disequilibrium.

Returning to Muller's argument that recombination should speed up evolution, our newfound knowledge of the evolutionary consequences of recombination has turned up a potential problem. Recombination not only brings alleles together, but also breaks down associations between alleles. This can actually retard the rate of evolution if the repulsion gametes, A_1B_2 and A_2B_1, are less fit than the A_1B_1 gamete. For an in-depth discussion of Muller's model and other aspects of the evolution of sex, read John Maynard Smith's *Evolutionary Genetics* (1989).

Muller's Ratchet

One advantage of sex is to speed up evolution through either segregation or recombination. In a world populated by evolving predators, prey, and pathogens and in a physical environment that is constantly changing, the ability to evolve quickly must enhance a species' long-term fitness appreciably. The origin and maintenance of sex may be attributed to its role in speeding up the rate of evolution. In addition, sex also plays a role in removing deleterious mutations from a population. H.J. Muller was the first to point this out by describing a phenomenon that has come to be called Muller's ratchet. Muller pointed out that, without sex, deleterious mutations may accumulate on chromosomes faster than selection can remove them from the population, leading to the ultimate extinction of the species. The operation of the ratchet involves three evolutionary forces: mutation, selection, and genetic drift, as will be described in this section.

Picture a population of N parthenogenetic diploid individuals. Each individual will have a certain number of deleterious mutations sprinkled around on its chromosomes. Assume that the mutation rate to deleterious alleles at any particular locus is so small that all deleterious mutations are heterozygous with the normal allele. The individuals may be grouped according to the number of deleterious mutations that they possess. A fraction x_0 of individuals will have no deleterious mutations, a fraction x_1 will have one deleterious mutation, and so forth. Note that the class of individuals with i deleterious mutations is genetically heterogeneous because different individuals are likely to have their i deleterious mutations at different loci.

If the number of individuals with no mutations, Nx_0, is small, then this class will be subject to the action of genetic drift, just as a rare mutation is subject to genetic drift. (Demographic stochasticity is the only source of randomness, as segregation does not occur in asexual organisms.) Should drift cause the loss of all individuals with no mutations, then each individual in the population will have at least one deleterious mutation, and Muller's ratchet will have clicked once. If the number of individuals with one mutation, Nx_1, is small, then this class will be subject to loss by genetic drift. If this class is lost, each individual will have at least two mutations, and the ratchet will have clicked once again. The rate of clicking of the ratchet is set by the time to loss of the smallest class by genetic drift. If the parameters are appropriate, then the species slowly accumulates deleterious mutations, leading to its eventual extinction.

For Muller's ratchet to work, the numbers of individuals in the class with the fewest mutations must be small. Population genetics can tell us the size of this class, if we are willing to make a few critical assumptions. The most important assumption is that the fitness of an individual with i deleterious mutations is

$$w_i = (1 - hs)^i.$$

The fitness contributions of the i heterozygous loci for deleterious mutations are multiplied together to obtain the fitness of the individual. The interaction of alleles at different loci is called *epistasis*; the form of epistasis used here is called *multiplicative epistasis*. Multiplicative epistasis implies that the effects on the probability of survival of individual loci are independent. If your chance of surviving from the effects of each of two loci in isolation is 1/2, then your chance of surviving their joint effects is 1/4. Later in this chapter we will consider another form of epistasis and some of the relevant experimental literature.

The derivation of the frequency of the classes of individuals with i mutations requires a knowledge of the Poisson distribution. Our basic strategy is to assume that the number of deleterious mutations in an individual is Poisson-distributed and then to find the mean of the Poisson as a function of the mutation rate and the strength of selection. Why a Poisson? You will have to wait until the derivation is finished; then you will see.

Assume, then, that the number of mutations per individual is Poisson-distributed with mean μ_K,

$$\text{Prob}\{i \text{ mutations}\} = \frac{e^{-\mu_K} \mu_K^i}{i!}.$$

With each reproduction, assume further that an offspring receives a Poisson-distributed number of new mutations with mean U,

$$\text{Prob}\{j \text{ new mutations}\} = \frac{e^{-U} U^j}{j!}.$$

The number of mutations after reproduction is the sum of two Poisson distributions, one with mean μ_K representing the number of mutations per individual before reproduction and one with mean U representing the number of new mutations. As the sum of two Poisson random variables is Poisson-distributed with mean equal to the sum of the means of the two Poissons, we see immediately that the mean number of mutations per individual after reproduction is Poisson with mean

$$\mu' = \mu_K + U.$$

Thus, the frequency of individuals with i mutations after reproduction is

$$x_i' = \frac{e^{-\mu'} \mu'^i}{i}.$$

The frequency of an individual with i mutations after selection is proportional to its frequency before selection times its fitness,

$$x_i'' = \frac{x_i'(1-hs)^i}{\sum_{j=0}^{\infty} x_j'(1-hs)^j}. \quad \text{...(7)}$$

The numerator is

$$\frac{e^{-\mu'} [\mu'(1-hs)]^i}{i}.$$

The denominator is

$$\sum_{j=0}^{\infty} x_j'(1-hs)^j = e^{-\mu'} \sum_{j=0}^{\infty} \frac{[\mu'(1-hs)]^j}{j!}$$

$$= e^{-\mu'} e^{\mu'(1-hs)}.$$

The third line is obtained from the Taylor series expansion of the exponential function,

$$e^x = \sum_{i=0}^{\infty} \frac{x^i}{i!}.$$

Thus, Equation 7 becomes

$$\frac{e^{-\mu'(1-hs)}[\mu'(1-hs)]^i}{i!},$$

which is, once again, a Poisson distribution. The mean of the new Poisson distribution is

$$\mu'_K = (\mu_K + U)(1 - hs).$$

At equilibrium, the mean does not change, so

$$\hat{\mu}_K = \frac{U(1-hs)}{hs} \approx \frac{U}{hs},$$

from which we obtain the frequency of individuals with i mutations at equilibrium,

$$\hat{x}_i \approx \frac{e^{-U/hs}(U/hs)^i}{i!}.$$

For Muller's ratchet to operate, the numbers of individuals with no mutations must be small enough that they will be eliminated in a reasonable time by genetic drift. Suppose, for example, that the total genomic mutation rate to deleterious alleles is $U = 1$, that the average selection against mutations in the heterozygous state is $hs = 0.1$, and that the population size is $N = 10^5$. In this case the number of individuals with no deleterious mutations is

$$Nx_0 = 4.54,$$

which is so small that genetic drift will soon eliminate them, even though they are the most fit genotypes in the population. However, if the population size is much larger, say $N = 10^7$, then $Nx_0 = 454$ and the ratchet comes to a grinding halt because the waiting time to fix the class of individuals with the fewest mutations becomes extraordinarily long. In those cases where the ratchet does operate, the accumulation of deleterious mutations proceeds at a rate determined by N, hs, and U. As there is no sex, hence no recombination, there is no way to reverse the steady reduction of fitness.

When there is sex, recombination can generate chromosomes with fewer (and more) mutations. Thus, the classes with fewer mutations are continually regenerated so that sex, with its accompanying recombination, may slow down or even reverse the clicking of Muller's ratchet. For this reason, sex is viewed as a way of eliminating deleterious mutations and hence is favoured by natural selection. Unfortunately, the story must end here because very little is known about the dynamics of deleterious mutations in finite populations with recombination and even less is known about the evolution of mutations promoting sex.

Because of the difficulty in obtaining a quantitative value for its advantage, it remains unclear whether Muller's ratchet can account for the maintenance of sex. However, the ratchet is often evoked with confidence to explain the loss of genes from chromosomes that do not recombine. For example, in many species in which the male is the heterogametic sex, the *Y* chromosome does not recombine with either the *X* chromosome or other *Y* chromosomes. Without recombination, the *Y* chromosome should slowly accumulate deleterious mutations and, quite plausibly, eventually fail to have any functional genes at all. In many species, including our own, the *Y* chromosome is almost entirely heterochromatic with only a few loci. Muller's ratchet is a perfectly viable explanation for this fact.

Kondrashov's Hatchet

The assumption of multiplicative epistasis in the model of Muller's ratchet is contradicted by a number of experimental studies. In this study, Mukai accumulated mutations on second chromosomes of *Drosophila melanogaster* for 60 generations under conditions that minimized natural selection. As the mutations accumulated, the relative viability of the second chromosomes when homozygous decreased. Mukai estimated that the mutation rate to deleterious mutations with measurable effects was 0.1411 per second chromosome per generation. For example, at 60 generations, a typical second chromosome would have $60 \times 0.1411 = 8.46$ mutations, which corresponds to the rightmost point on the graph.

The three curves on the graph correspond to three different models of epistasis. The lower convex curve is for multiplicative epistasis of the form

$$w_n = (1 - s)^n,$$

where n is the number of deleterious mutations, s is a measure of the effect of each mutation, and w_n is the relative viability of a fly with n homozygous deleterious mutations. The method of least squares was used to estimate s, whose value is 0.05897. The middle straight line corresponds to additive epistasis

$$w_n = 1 - sn$$

with $s = 0.04749$. Finally, the upper concave curve is for a quadratic model of synergistic epistasis,

$$w_n = 1 - sn - an^2,$$

where $s = 0.009813$ and $a = 0.00555$. The figure leaves little doubt that synergistic epistasis fits the data much better than do additive or multiplicative epistasis. With synergistic epistasis, fitness drops off faster than expected from the effects of mutations in isolation. It is an embodiment of the notion of "things going from bad to worse."

Synergistic epistasis does not invalidate Muller's ratchet, but it does change its rate. Unfortunately, the multiplicative assumption is directly responsible for the Poisson distribution of the number of mutations per genome. Changing this assumption necessitates a much more difficult mathematical development. Rather than pursuing that gruesome prospect, a much more exciting direction is to investigate the effects of synergistic epistasis itself on the fate of sexual and asexual species. This will lead to another argument for the evolutionary advantage of sex based on the ability of sexual species to eliminate deleterious mutations more effectively than do asexual species.

The importance of synergistic epistasis for the maintenance of sex was emphasized in a paper by Alexey Kondrashov (1988), whose model was dubbed Kondrashov's hatchet by Michael Turelli, who is never wanting for a clever turn of phrase. Models of synergistic epistasis are very difficult to analyze. However, Kondrashov realized that truncation selection, which is well understood by quantitative geneticists, is really an instance of synergistic epistasis. Imagine that "deleterious" mutations accumulate in a genome with no effect on fitness until the number of mutations exceeds some magic number, k, at which point any additional mutations are lethal. Said another way: Fitness, as a function of the number of deleterious mutations, is equal to one until the number of mutations is greater than k; when the number of deleterious mutations is greater than k, fitness is zero. The fitness could be written as

$$w_n = \begin{cases} 1 & \text{if } n \le k \\ 0 & \text{if } n > k \end{cases}.$$

The hatchet falls on all genotypes with more than k mutations.

If the number of mutations is now viewed as a quantitative trait, then Kondrashov's hatchet is formally the same as truncation selection. Accordingly, we will use our results on the response to selection to study the evolution of the number of deleterious mutations per individual in sexual organisms. But first, we will record some observations about selection in asexual organisms to see the relative advantages of being sexual.

Mutation is added to Kondrashov's model in the same way as was done for Muller's ratchet. Each generation, every offspring receives a Poisson-distributed number of new deleterious mutations. The mean of the Poisson is U. Through time, the number of deleterious mutations in any lineage will increase until it reaches k. From then on, any offspring with additional mutations will die. After a sufficient period of time, every individual in the population will have exactly k mutations. From this point on, the frequency of offspring in the population with no additional mutations is the probability that a Poisson random variable with mean U is zero,

$$\text{Prob}\{X = 0\} = e^{-U}.$$

The probability that an offspring has one or more new mutations is $1 - e^{-U}$. The frequency of individuals with $k = 10$ or more mutations, before selection and when $U = 1$, is illustrated.

The genetic load of the equilibrium asexual population is just

$$\frac{w_{max} - \bar{w}}{w_{max}} = 1 - [e^{-U} \times 1 + (1 - e^{-U}) \times 0]$$

$$= 1 - e^{-U}.$$

Once again, we have a calculation that shows that the genetic load depends only on the mutation rate and is independent of the strength of selection. For small U, the load is approximately

$$1 - (1 - U) = U,$$

which is the same. There, the contribution of a single locus to the load was found to be $2u$, where u is the mutation rate per allele. Here, U is the mutation rate per diploid genome, which is twice the rate per haploid genome. This difference accounts for the factor of two.

As an aside, it is worth noting that the equilibrium mutational load for an asexual species with any sort of epistasis is remarkably easy to derive by considering only the frequency of the genomes that are free of deleterious mutations, x_0. In the next generation, the frequency of this class is

$$x_0' = \frac{x_0 w_0 e^{-U}}{\bar{w}}.$$

That is, its frequency is equal to its frequency in the current generation times the probability that an offspring gets no additional mutations (e^{-U} is the probability that the Poisson random variable is zero) times the fitness of the deleterious-free class, w_0, which is one in our models, divided by the mean fitness of the population. At equilibrium, $x_0' = x_0$, so

$$\bar{w} = e^{-U}.$$

The simplicity of the result might be unsettling at first, but really is as simple as it appears.

Consider now the situation in a sexual species. As a first step, imagine what would happen if the asexual species were suddenly to turn sexual. With meiosis and crossing over, individuals would appear with fewer than k mutations. They, along with those with k mutations, would live, while those with more than k would die. As a result, the mean number of deleterious alleles would be less than in the asexual population. With sex, selection lowers the mean number of mutations per individual, while mutation increases the number. Eventually an equilibrium is reached where the genetic load must be lower than the asexual load. By how much lower? Read on!

Kondrashov's hatchet kills off all offspring with more than k mutations, which reduces the average number of deleterious mutations per individual by an amount that is, by definition, the selection differential, S. As there is no environmental contribution to the number of deleterious mutations, the heritability of this phenotype is $h^2 = 1$. The response to selection is

$$R = h^2 S = S.$$

Thus, each generation the average number of mutations per individual is increased by U and decreased by $R = S$. Eventually, an equilibrium will be reached where

$$U = S. \qquad \text{...(6.8)}$$

Our job is to discover the mutational load at this equilibrium and to see if it is sufficiently smaller than the asexual load to overcome the twofold cost of sex.

Recall that the selection intensity is

$$i(p) = \frac{S}{\sqrt{V_P}}.$$

Dividing both sides of Equation 8 by $\sqrt{V_P}$, we find that the selection intensity at equilibrium is

$$i(p) = \frac{U}{\sqrt{V_P}}.$$

The rightmost term, the ratio of the mean number of deleterious mutations in an offspring divided by the standard deviation of the total number of deleterious mutations per individual, is subject to experimental investigation. In *Drosophila*, Kondrashov argued that $U = 2$ and $\sqrt{V_P} = 10$, in which case the intensity of selection is $i(p) = 0.2$. We see that this intensity occurs when about 10 percent of the population is killed each generation by Kondrashov's hatchet. This is the mutational load of a sexual population at equilibrium. The genetic load of an asexual species is $1 - e^{-U} = 0.86$, much greater than the genetic load of sexual species (0.1). The ratio of the sexual to asexual mean fitnesses is about $0.9/0.14 = 6.4$, which easily overcomes the twofold cost of sex.

In this chapter, we have examined four possible explanations for the maintenance of sex in face of its twofold cost. Two of the explanations, those of Kirkpatrick and Jenkins and of Kondrashov, can be developed to the point of showing quantitatively that sex can plausibly achieve a twofold advantage. We have not addressed the actual evolution of sex. That is, we have not followed the fate of a mutation that confers sexuality on its bearer to see that sex will, in fact, evolve. Models of the evolution of sex are very difficult to analyze.

4

Equilibrium and Mating Systems

Evolution is a process that takes place in populations of organisms. To study evolution, we need to shift our focus to population genetics, the algebraic description of the genetic makeup of a population and the changes in allelic frequencies in populations over time. This chapter is the first of three that looks at what population genetics can tell us about the way evolution proceeds.

Almost all of the mathematical foundations of genetic changes in populations were developed in a short period of time during the 1920s and 1930s by three men: R.A. Fisher, J.B.S. Haldane, and S. Wright. Some measure of disagreement emerged among these men, but they disagreed on which evolutionary processes were more important, not on how the processes worked. Since the 1960s, excitement has arisen in the field of population genetics, primarily on three fronts. First, the high-speed computer has made it possible to do a large amount of arithmetic in a very short period of time; thus, complex simulations of real populations can be added to the repertoire of the experimental geneticist. Second, electrophoresis has provided a means of gathering the large amount of empirical data necessary to check some of the assumptions used in mathematical models. The information and interpretation of the electrophoretic data have generated some controversy about the role of "neutral" evolutionary changes in natural populations. Last, newer techniques of molecular genetics are being used to analyze the relationships among species and the rate of evolutionary processes. We consider these studies later.

Hardy-Weinberg Equilibrium

Let us begin with a few definitions. For the most part, we define a species as a group of organisms potentially capable of interbreeding. Most species are made up of populations, interbreeding groups of organisms that are usually subdivided into partially isolated breeding groups called *demes*. As we will see, it is these demes, or local populations, that can evolve.

In 1908, G. H. Hardy, a British mathematician, and W. Weinberg, a German physician, independently discovered a rule that relates allelic and genotypic frequencies in a population of diploid, sexually reproducing individuals if that population has random mating, large size, no mutation or migration, and no selection. The rule has three aspects:

1. The allelic frequencies at an autosomal locus in a population will not change from one generation to the next (allelic-frequency equilibrium).
2. The genotypic frequencies of the population are determined in a predictable way by the allelic frequencies (genotypic-frequency equilibrium).
3. The equilibrium is neutral. That is, if it is perturbed, it will be reestablished within one generation of random mating at the new allelic frequencies (if all the other requirements are maintained).

Calculating Allelic Frequencies

If we consider an autosomal locus in a diploid, sexually reproducing species, allelic frequencies can be measured in either of two ways. The first is simply by counting genes:

$$\text{frequency of the } a \text{ allele, } q, = \frac{\text{number of } a \text{ alleles}}{\text{total number of alleles}}$$

The expression "frequency of" can be shortened to *f*(). For example, the frequency of the a allele is written as *f*(*a*). Since the homozygotes have two of a given allele and heterozygotes have only one, and since the total number of alleles is twice the number of individuals (each individual carries two alleles), we can calculate allelic frequencies in the following manner. Consider, for example, the phenotypic distribution of MN blood types (controlled by the codominant *M* and *N* alleles) among two hundred persons chosen randomly in Columbus, Ohio:

$$\begin{aligned} \text{type M } (MM \text{ genotype}) &= 114 \\ \text{type MN } (MN \text{ genotype}) &= 76 \\ \text{type N } (NN \text{ genotype}) &= \underline{\ 10\ } \\ &\ \ \ \ 200 \end{aligned}$$

Then,

$$p = f(M) = \frac{2(114) + 76}{2(200)} = \frac{304}{400} = 0.76$$

$$q = f(N) = \frac{2(10) + 76}{2(200)} = \frac{96}{400} = 0.24$$

Alternatively, because the frequencies of the two alleles *M* and *N* must add up to unity ($p + q = 1$), $q = 1 - p$ (and $p = 1 - q$).

Another way of calculating allelic frequencies is based on knowledge of the genotypic frequencies, which in this example are:

$$f(MM) = \frac{114}{200} = 0.57$$

$$f(MN) = \frac{76}{200} = 0.38$$

$$f(NN) = \frac{10}{200} = 0.05$$

We derive an expression for calculating *p* and *q* based on genotypic frequencies as follows:

$$p = f(M) = \frac{2 \times \text{number of } MM + \text{number of } MN}{2 \times \text{total number}}$$

$$= \frac{2 \times \text{number of } MM}{2 \times \text{total number}} + \frac{\text{number of } MN}{2 \times \text{total number}}$$

$$= f(MM) + (1/2) f(MN)$$

and

$$q = f(N) = \frac{2 \times \text{number of } NN + \text{number of } MN}{2 \times \text{total number}}$$

$$= \frac{2 \times \text{number of } NN}{2 \times \text{total number}} + \frac{\text{number of } MN}{2 \times \text{total number}}$$

$$= f(NN) + (1/2) f(MN)$$

Thus allelic frequencies can be calculated as the frequency of homozygotes plus half the frequency of heterozygotes as follows:

$$p = f(M) = f(MM) + (1/2) f(MN)$$
$$= 0.57 + (1/2)0.38 = 0.76$$

$$q = f(N) = f(NN) + (1/2) f(MN)$$
$$= 0.05 + (1/2)0.38 = 0.24$$

or

$$q = 1 - p = 1 - 0.76 = 0.24.$$

Note that these two methods (counting alleles and using genotypic frequencies) are algebraically identical and thus give identical results.

Assumptions of Hardy-Weinberg Equilibrium

We will consider a population of diploid, sexually reproducing organisms with a single autosomal locus segregating two alleles (i.e., every individual is one of three genotypes—*MM*, *MN*, or *NN*). Later on, we generalize the discussion to include multiple alleles and multiple loci. For the moment, the focus is on a genetic system such as the *MN* locus in human beings. The following major assumptions are necessary for the Hardy-Weinberg equilibrium to hold.

Random mating

The first assumption is *random mating*, which means that the probability that two genotypes will mate is the product of the frequencies (or probabilities) of the genotypes in the population. If the *MM* genotype makes up 90% of a population, then any individual has a 90% chance (probability 0.9) of mating with a person with an *MM* genotype. The probability of an *MM* by *MM* mating is (0.9)(0.9), or 0.81.

Deviations from random mating come about for two reasons: choice or circumstance. If members of a population choose individuals of a particular phenotype as mates more or less often than at random, the population is engaged in *assortative mating*. If individuals with similar phenotypes are mating more often than at random, *positive assortative mating* is in force; if matings occur between individuals with dissimilar phenotypes more often than at random, *negative assortative mating*, or *disassortative mating*, is at work.

Deviations from random mating also arise when mating individuals are either more closely related genetically or more distantly related than individuals chosen at random from the population. *Inbreeding* is the mating of related individuals, and *outbreeding* is the mating of genetically unrelated individuals. Inbreeding is a consequence of pedigree relatedness (e.g., cousins) and small population size.

One of the first counterintuitive observations of population genetics is that deviations from random mating alter genotypic frequencies but not allelic frequencies. Envision a population in which every individual is the parent of two children. On the average, each individual will pass on one copy of each of his or her alleles. Assortative mating and inbreeding will change the zygotic (genotypic) combinations from one generation to the next, but will not change which alleles are passed into the next generation. Thus genotypic, but not allelic, frequencies change under nonrandom mating.

Large population size

Even when an extremely large number of gametes is produced in each generation, each successive generation is the result of a sampling of a relatively small portion of the gametes of the previous generation. A sample may not be an accurate representation of a population, especially if the sample is small. Thus, the second assumption of the Hardy-Weinberg equilibrium is that the population is infinitely large. A large population produces a large sample of successful gametes. The larger the sample, the greater the probability that the allelic frequencies of the offspring will accurately represent the allelic frequencies in the parental population. When populations are small or when alleles are rare, changes in allelic frequencies take place due to chance alone. These changes are referred to as *random genetic drift*, or just *genetic drift*.

No mutation or migration

Allelic and genotypic frequencies may change through the loss or addition of alleles through mutation or migration (immigration or emigration) of individuals from or into a population. The third and fourth assumptions of the Hardy-Weinberg equilibrium are that neither mutation nor migration causes such allelic loss or addition in the population.

No natural selection

The final assumption necessary to the Hardy-Weinberg equilibrium is that no individual will have a reproductive advantage over another individual because of its genotype. In other words, no natural selection is occurring. (Artificial selection, as practiced by animal and plant breeders, will also perturb the Hardy-Weinberg equilibrium of captive populations.)

In summary, the Hardy-Weinberg equilibrium holds (is exactly true) for an infinitely large, randomly mating population in which mutation, migration, and natural selection do not occur. In view of these assumptions, it seems that such an equilibrium would never be characteristic of natural populations. However, this is not the case. Hardy-Weinberg equilibrium is approximated in natural populations for two major reasons. First, the consequences of violating some of the assumptions, such as no mutation or infinitely large population size, are small. Mutation rates, for example, are on the order of one change per locus per generation per 106 gametes. Thus, there is virtually no measurable effect of mutation in a single generation. In addition, populations do not have to be infinitely large to act as if they were.

As we will see, a relatively small population can still closely approximate Hardy-Weinberg equilibrium. In other words, minor deviations from the other assumptions can still result in a good fit to the equilibrium; only major deviations can be detected statistically. Second, the Hardy-Weinberg equilibrium is extremely resilient to change because, regardless of the perturbation, the equilibrium is usually reestablished after only one generation of random mating. The new equilibrium will be, however, at the new allelic frequencies—the Hardy-Weinberg equilibrium does not "return" to previous allelic values.

Proof of Hardy-Weinberg Equilibrium

The three properties of the Hardy-Weinberg equilibrium are that (1) allelic frequencies do not change from generation to generation, (2) allelic frequencies determine genotypic frequencies, and (3) the equilibrium is achieved in one generation of random mating. We will concentrate for a moment on the second property. In a population of individuals segregating the *A* and *a* alleles at the A locus, each individual will be one of three genotypes: *AA*, *Aa*, or *aa*. If p $f(A)$ and q $f(a)$, then we can predict the genotypic frequencies in the next generation. If all the assumptions of the Hardy-Weinberg equilibrium are met, the three genotypes should occur in the population in the same frequencies at which gametes would be randomly drawn in pairs from a *gene pool*. A gene pool is defined as all of the alleles available among the reproductive members of a population from which gametes can be drawn. Thus,

$$f(AA) = (p \times p) = p^2$$
$$f(Aa) = (p \times q) + (q \times p) = 2pq$$
$$f(aa) = (q \times q) = q^2$$

demonstrates the second property of the Hardy-Weinberg equilibrium.

Another way of demonstrating the properties of the Hardy-Weinberg equilibrium for the one-locus, two-allele case in sexually reproducing diploids is by simply observing the offspring of a randomly mating, infinitely large population. Let the initial frequencies of the three genotypes be any values that sum to one; for example, let *X*, *Y*, and *Z* be the proportions of the *AA*, *Aa*, and aa genotypes, respectively. The proportions of offspring after one generation of random mating are shown. For example, the probability that an *AA* individual will mate with an *AA* individual is $X \times X$, or X^2. Since all the offspring of this mating are *AA*, they are counted only under the *AA* column of offspring. When all possible matings are counted, the offspring with each genotype are summed. The proportion of AA offspring is $X^2 + XY + (1/4)Y^2$, which factors to $(X + [1/2]Y)^2$. Recall that the frequency of an allele is the frequency of its homozygote plus half the frequency of the heterozygote. Hence, $X + (1/2)Y$ is the frequency of A, since $X = f(AA)$ and $Y = f(Aa)$. If $p = f(A)$, then $(X + [1/2]Y)^2$ is p^2. Thus, after one generation of random mating, the proportion of AA homozygotes is p^2. Similarly, the frequency of *aa* homozygotes after one generation of random mating is $Z^2 + YZ + (1/4)Y^2$, which factors to $(Z + [1/2]Y)^2$, or q^2. The frequency of heterozygotes when summed and factored is $2(X + [1/2]Y)(Z + [1/2]Y)$, or $2pq$. Therefore, after one generation of random mating, the three genotypes (*AA*, *Aa*, and *aa*) occur as p^2, $2pq$, and q^2.

Looking at the first property of the Hardy-Weinberg equilibrium, that allelic frequencies do not change generation after generation, we can ask, Have the allelic frequencies changed from one generation to the next (from the parents to the offspring)? Before random mating, the frequency of the A allele is, by definition, p:

$$f(A) = p = f(AA) + (1/2)f(Aa) = X + (1/2)Y$$

After random mating, the frequency of the A homozygote is p^2, and the frequency of the heterozygote is $2pq$. Thus, the frequency of the A allele, the frequency of its homozygote plus half the frequency of the heterozygotes, is

$$\begin{aligned} f(A) &= f(AA) + (1/2)f(Aa) \\ &= p^2 + (1/2)(2pq) \\ &= p^2 + pq = p(p+q) \\ &= p (\text{remember, } p + q = 1) \end{aligned}$$

Thus, in a randomly mating population of sexually reproducing diploid individuals, the allelic frequency, p, does not change from generation to generation. Here, by observing the offspring of a randomly mating population, we have proven all three properties of the Hardy-Weinberg equilibrium.

Generation Time

Although generation interval is commonly thought of as the average age of the parents when their offspring are born, the statistical concept of a generation is more complex. Demographers use formulas that relate generation time to the age of reproducing females, the reproductive level of each age group, and the probability of survival in each age group. Here, to avoid these complexities, we will use *discrete generations*, unless otherwise noted. That is, we will assume that all the individuals drawn in a sample, for purposes of determining allelic and genotypic frequencies, are drawn from the same generation, and that, in resampling the population, the second sample represents the offspring of the first generation. The discrete-generation model holds for organisms such as annual plants and fruit flies maintained under laboratory conditions, with no breeding among individuals of different generations. Generations that overlap, as in populations of human beings and many other organisms, usually are better described by somewhat more complex mathematical models.

Testing for Fit to Hardy-Weinberg Equilibrium

There are several ways to determine whether a given population conforms to the Hardy-Weinberg equilibrium at a particular locus. However, the question usually arises when there is just a single sample from a population, representing only one generation. Can the existence of the Hardy-Weinberg equilibrium be determined with just one sample? The answer is that we can determine whether the three genotypes (*AA*, *Aa*, and *aa*) occur with the frequencies p^2, $2pq$, and q^2. If they do, then the population is considered to be in Hardy-Weinberg proportions; if not, then the population is not considered to be in Hardy-Weinberg proportions.

MN blood types

To determine whether observed and expected allelic frequencies are the same, we can use the chi-square statistical test. In a chi-square test, we compare an observed number with an expected number. In this case, the observed values are the actual numbers of the three genotypes in the sample, and the expected values come from the prediction that the genotypes will occur in the p^2, $2pq$, and q^2 proportions. The agreement between observed and expected numbers is very good, obvious even before the calculation of the chi-square value. Since the critical chi-square for one degree of freedom at the 0.05 level is 3.841, we find that the Ohio population does not deviate from Hardy-Weinberg proportions at the MN locus.

Earlier, we used the chi-square statistic to test how well real data fit an expected data set based on a ratio predicted before the test. In that case, the number of degrees of freedom was simply the number of independent categories: the total number of categories minus one. Here, however, our expected ratio is derived from the data set itself. The values p^2, $2pq$, and q^2 came from p and q, which were estimated from the data. In this case, we lose one additional degree of freedom for every independent value we estimate from the data. If we calculate p from a sample, we lose one degree of freedom. However, we do not lose a degree of freedom for estimating q, since q is no longer an independent variable: $q = 1 - p$. So in the previous case, we lose two degrees of freedom—one for estimating p and one for independent categories. The general rule of thumb in using chi-square analysis to test for data fit to Hardy-Weinberg proportions is that the number of degrees of freedom must equal the number of phenotypes minus the number of alleles (in this case, $3 - 2 = 1$).

The chi-square may seem paradoxical. Because the observed allelic frequencies calculated from the original genotypic data are used to calculate the expected genotypic frequencies, it may appear to some individuals that the analysis must, by its very nature, show that the population is in Hardy-Weinberg proportions. We use data similar to the Ohio sample, except that the original number of heterozygotes has been distributed equally among the two homozygote classes. The same allelic frequencies are maintained, yet the genotypic distribution differs. The chi-square value of 200.00 for these data demonstrates that the population is not in Hardy-Weinberg proportions. Thus, a chi-square analysis of fit to the Hardy-Weinberg proportions by no means represents circular reasoning.

PKU

Circumstances sometimes do not allow us to test for Hardy-Weinberg proportions. In the case of a dominant trait, for example, allelic frequencies cannot be calculated from the genotypic classes because the homozygous dominant individuals cannot be distinguished from the heterozygotes. However, we can estimate allelic frequencies by assuming that the Hardy-Weinberg equilibrium exists and, thereby, assuming that the frequency of the recessive homozygote is q^2, from which q and then p can be estimated.

If, for example, Hardy-Weinberg equilibrium is assumed for a disease such as phenylketonuria (PKU), which is expressed only in the homozygous recessive state, it is possible to calculate the proportion of the population that is heterozygous (carriers of the PKU allele). But is it fair to assume Hardy-Weinberg equilibrium here? Until recent medical advances allowed intervention, there was a good deal of selection against individuals with PKU, who were usually mentally retarded. Thus the assumption of no selection, required for equilibrium, is violated. However, only one child in ten thousand live births has PKU. When a genotype is as rare as one in ten thousand, selection has a negligible effect on allelic frequencies. Therefore, because of the rarity of the trait, we can assume Hardy-Weinberg equilibrium and calculate

frequency of recessive homozygote = q^2 = 1/10,000 = 0.0001

so,

$$q = \sqrt{0.0001} = 0.01$$

and

$$p = 1 - q = 0.99$$

Therefore,

frequency of normal homozygote = p^2 = $(0.99)^2$ = 0.98 or 98 in 100

frequency of heterozygote = $2pq$ = 2(0.01)(0.99) = 0.02 or in 100.

By assuming the Hardy-Weinberg equilibrium, we have discovered something not intuitively obvious: A recessive gene causing a trait as rare as one in ten thousand is carried in the heterozygous state by one individual in fifty. Obviously, the chi-square test cannot be used to verify the Hardy-Weinberg proportions since we derived the allelic frequencies by assuming Hardy-Weinberg proportions to begin with. In statistical terms, the number of phenotypes minus the number of alleles = 2 – 2 = 0 degrees of freedom, which precludes doing a chi-square test.

Extensions of Hardy-Weinberg Equilibrium

The Hardy-Weinberg equilibrium can be extended to include, among other cases, multiple alleles and multiple loci.

Multiple Alleles

Multinomial expansion

The expected genotypic array under Hardy-Weinberg equilibrium is p^2, $2pq$, and q^2, which form the terms of the binomial expansion $(p + q)^2$. If males and females each have the same two alleles in the proportions of p and q, then genotypes will be distributed as a binomial expansion in the frequencies p^2, $2pq$, and q^2. To generalize to more than two alleles, one need only add terms to the binomial expansion and thus create a multinomial expansion. For example, with alleles a, b, and c with frequencies p, q, and r, the genotypic distribution should be $(p + q + r)^2$, or

$$p^2 + 2pq + 2pr + q^2 + 2qr + r^2$$

Homozygotes will occur with frequencies p^2, q^2, and r^2, and heterozygotes will occur with frequencies $2pq$, $2pr$, and $2qr$. The ABO blood-type locus in human beings is an interesting example because it has multiple alleles and dominance.

ABO blood groups

The ABO locus has three alleles: I^A, I^B, and i, with the I^A and I^B alleles codominant, and both dominant to the i allele. These alleles control the production of a surface antigen on red blood cells. Table contains blood-type data from a sample of five hundred persons from Massachusetts. Is the population in Hardy-Weinberg proportions? The answer is not apparent from the data alone, since there are two possible genotypes for both the A and the B phenotypes. No estimate of the allelic frequencies is possible without making assumptions about the number of each genotype within these two phenotypic classes. Is it possible to estimate the allelic frequencies? The answer is yes, if we assume that Hardy-Weinberg equilibrium exists.

One procedure follows. Let us assume that $p = f(I^A)$, $q = f(I^B)$, and $r = f(i)$. Blood type O has the ii genotype; if the population is in Hardy-Weinberg proportions, this genotype should occur at a frequency of r^2. Thus

$$f(ii) = 231/500 = 0.462 = r^2$$

and

$$r = f(i) = \sqrt{0.462} = 0.680$$

We see that blood type A plus blood type O include only the genotypes $I^A I^B$, I^Ai, and ii. If the population is in Hardy-Weinberg proportions, these together should be $(p + r)^2$, in which $p^2 = f(I^AI^A)$, $2pr = f(I^Ai)$, and $r^2 = f(ii)$:

$$(p+r)^2 = (199+231)/500 = 0.860$$

Then, taking the square root of each side

$$p+r = \sqrt{0.860} = 0.927$$

and

$$p = 0.927 - r = 0.927 - 0.680 = 0.247$$

The frequency of allele I^B, q, can be obtained by similar logic with blood types B and O, or simply by subtraction:

$$q = 1-(p+r) = 1-0.927 = 0.073$$

Thus, the Hardy-Weinberg equilibrium can be extended to include multiple alleles and can be used to make estimates of the allelic frequencies in the ABO blood groups. With ABO, it is statistically feasible to do a chi-square test because there is one degree of freedom (number of phenotypes number of alleles 4 – 3 = 1). We are really testing only the AB and B categories; if we did our calculations as shown, the observed and expected values of phenotypes A and O must be equal.

Multiple loci

The Hardy-Weinberg equilibrium can also be extended to consider several loci at the same time in the same population. This situation deserves mention because the whole genome is likely involved in evolutionary processes and we must, eventually, consider simultaneous allelic changes in all loci segregating alleles in an organism. (Even with a high-speed computer, simultaneous consideration of many loci is a bit far off in the future.) When two loci, *A* and *B*, on the same chromosome are in equilibrium with each other, the combinations of alleles on a chromosome in a gamete follow the product rule of probability. Consider the *A* locus with alleles *A* and *a* and the *B* locus with alleles *B* and *b*, respectively, with allelic frequencies p_A and q_A for *A* and *a*, respectively, and p_B and q_B for *B* and *b*, respectively. Given completely random circumstances, the chromosome with the *A* and *B* alleles should occur at the frequency $p_A\, p_B$. This is referred to as *linkage equilibrium*. When alleles of different loci are not in equilibrium (i.e., not randomly distributed in gametes), the condition is referred to as *linkage disequilibrium*. The approach to linkage equilibrium is gradual and is a function of the recombination distance between the two loci.

For example, let's start with a population out of equilibrium so that all chromosomes are *AB* (70%) or *ab* (30%). Then $p_A = 0.7$, $q_A = 0.3$, $p_B = 0.7$, and $q_B = 0.3$. We expect the *Ab* chromosome to occur $0.7 \times 0.3 = 0.21$, or 21% of the time. The frequency of the *Ab* chromosome is zero. Assume the map distance between the two loci is 0.1; in other words, 10% of chromatids in gametes are recombinant. Initially, we consider that each locus is in Hardy-Weinberg proportions, or the frequency of *AB/AB* individuals = 0.49 (0.7×0.7); the frequency of *ab/ab* individuals is 0.09 (0.3×0.3); and the frequency of *AB/ab* individuals is 0.42 ($2 \times 0.7 \times 0.3$).

After one generation of random mating, gametes will be as follows:
from *AB/AB* individuals (49%): only *AB* gametes, 49% of total
from *ab/ab* individuals (9%): only *ab* gametes, 9% of total
from *AB/ab* individuals (42%):
AB gametes, 18.9% of total (0.45×0.42)
ab gametes, 18.9% of total (0.45×0.42)

Ab gametes, 2.1% of total (0.05 × 0.42)

aB gametes, 2.1% of total (0.05 × 0.42)

(The values of 18.9% and 2.1% for the dihybrids result from the fact that since map distance is 0.1, 10% of gametes will be recombinant, split equally between the two recombinant classes—5% and 5%. Ninety percent will be parental, split equally between the two parental classes—45% and 45%. Each of these numbers must be multiplied by 0.42 because the dihybrid makes up 42% of the total number of individuals.)

Although we expect 21% of the chromosomes to be of the *Ab* type, only 2.1%, 10% of the expected, appear in the gene pool after one generation of random mating. You can see that linkage equilibrium is achieved at a rate dependent on the map distance between loci. Unlinked genes, appearing 50 map units apart, also gradually approach linkage equilibrium.

Although we will not derive these extensions here, we note two others. If the frequencies of alleles at an autosomal locus differ in the two sexes, it takes two generations of random mating to achieve equilibrium. In the first generation, the allelic frequencies in the two sexes are averaged so that each sex now has the same allelic frequencies. Genotypic frequencies then come into Hardy-Weinberg proportions in the second generation. However, if the allelic frequencies differ in the two sexes for a sex-linked locus, Hardy-Weinberg proportions are established only gradually. The reasoning is straightforward. Females, with an X chromosome from each parent, average the allelic frequencies from the previous generation. However, males, who get their X chromosomes from their mothers, have the allelic frequencies of the females in the previous generation. Hence, the allelic frequencies are not the same in the two sexes after one generation of random mating, and equilibrium is achieved slowly.

Nonrandom Mating

The Hardy-Weinberg equilibrium is based on the assumption of random mating. Deviations from random mating come about when phenotypic resemblance or relatedness influences mate choice. When phenotypic resemblance influences mate choice, either *assortative* or *disassortative* mating occurs, depending on whether individuals choose mates on the basis of similarity or dissimilarity, respectively. For example, in human beings, assortative mating occurs for height—short men tend to marry short women, and tall men tend to marry tall women. When relatedness influences mate choice, either *inbreeding* or *outbreeding* occurs, depending on whether mates are more or less related than two randomly chosen individuals from the population. An example of inbreeding in human beings is marriage between first cousins. Both types of nonrandom mating (assortative-disassortative mating and inbreeding-outbreeding) have the same qualitative effects on the Hardy-Weinberg equilibrium: assortative mating and inbreeding increase homozygosity without changing allelic frequencies, whereas disassortative mating and outbreeding increase heterozygosity without changing allelic frequencies.

Two differences are apparent, however, between the effects of phenotypic resemblance and relatedness on mate choice. First, assortative or disassortative mating disturbs the Hardy-Weinberg equilibrium only when the phenotype and genotype are closely related. That is, if assortative mating occurs for a nongenetic trait, then the Hardy-Weinberg equilibrium will not be distorted. Inbreeding and outbreeding affect the genome directly. A second difference between the two types of mating is that the effects of inbreeding or outbreeding are felt across the whole genome, whereas the disturbances to the equilibrium caused by assortative and disassortative mating occur only for the particular trait being considered (and for closely linked loci). Given the similarities in the consequences of the two types of matings, we will concentrate our discussion on inbreeding.

Inbreeding

Inbreeding comes about in two ways: (1) the systematic choice of relatives as mates and (2) the subdivision of a population into small subunits, leaving individuals little choice but to mate with relatives. We will concentrate on inbreeding as the systematic choice of relatives as mates. The consequences of both are similar.

Common ancestry

An inbred individual is one whose parents are related—that is, there is *common ancestry* in the family tree. The extent of inbreeding thus depends on the degree of common ancestry that the parents of an inbred individual share. When mates share ancestral genes, each may pass on copies of the same ancestral allele to their offspring. An inbred individual can then carry identical copies of a single ancestral allele. In other words, an individual of *aa* genotype is homozygous and, if it is possible that the *a* allele from each parent is a length of DNA originally copied from a common ancestor, the *aa* individual is said to be inbred.

The first observable effect of inbreeding is the expression of hidden recessives. In human beings, each individual carries, on the average, about four *lethal-equivalent alleles*, alleles that kill when paired to form a homozygous genotype. In many, and probably most, human societies, zygotes are generally heterozygous for these lethal alleles because of a cultural pattern of outbreeding, mating with nonrelatives. Rarely does an outbred zygote receive the same recessive lethal from each parent. Dominance acts to mask the expression of deleterious recessive alleles. But, in the process of inbreeding, when the zygote may receive copies of the same ancestral allele from each parent, there is a substantial increase in the probability that a deleterious allele will pair to form a homozygous genotype. Inbreeding can result in spontaneous abortions (miscarriages), fetal deaths, and congenital deformities. In many species, however, inbreeding—even self-fertilization— occurs normally. These species usually do not have the problem with lethal equivalents that species that normally outbreed do. Through time, species that normally inbreed have had these deleterious alleles mostly eliminated, presumably by natural selection. Inbreeding has even been used successfully for artificial selection in live-stock and crop plants.

From our previous discussion, you can see that there are two types of homozygosity—*allozygosity*, in which two alleles are alike but unrelated (not copies of the same ancestral allele) and *autozygosity*, in which two alleles have *identity by descent* (i.e., are copies of the same ancestral allele). An *inbreeding coefficient*, F, can be defined as the probability of autozygosity, or the probability that the two alleles in an individual at a given locus are identical by descent. This coefficient can range from zero, at which point there is no inbreeding, to one, at which point it is certain an individual is autozygous.

Increased homozygosity from inbreeding

What are the effects of inbreeding on the Hardy-Weinberg equilibrium? Let us for a moment return to the gene pool concept to produce zygotes. Assume that an allele drawn from this gene pool is of the A type, drawn with a probability of p. On the second draw, the probability of autozygosity, that is, of drawing a copy of the same allele A, is F, the inbreeding coefficient. Thus the probability of an autozygous AA individual is pF. On the second draw, however, with probability $(1 - F)$, either the A or a allele can be drawn, with probabilities of $p^2(1 - F)$ and $pq(1 - F)$, respectively. Note that a second A allele produces a homozygote that is not inbred (allozygous).

If the first allele drawn was an *a* allele, with probability q, then the probability of drawing the same allele (copy of the same ancestral allele) is F, and thus the probability of autozygosity is qF. However, the probability of drawing an *a* or *A* allele that does not contribute to inbreeding is $(1 - F)$ and, therefore, the probability of an *aa* or *Aa* genotype is $q^2(1 - F)$ and $pq(1 - F)$, respectively. These calculations are a summary of the genotypic proportions in a population with inbreeding.

First, when the inbreeding coefficient is zero (completely random mating), the table reduces to Hardy-Weinberg proportions. Second, compared with Hardy-Weinberg proportions, inbreeding increases the proportion of homozygotes in the population (identity by descent implies homozygosity). With complete inbreeding ($F = 1$), only homozygotes will occur in the population.

How does inbreeding affect allelic frequencies? Recall that an allelic frequency is calculated as the frequency of homozygotes for one allele plus half the frequency of the heterozygotes. Here we let p_{n+1} be the frequency of the *A* allele after one generation of inbreeding:

$$\begin{aligned} p_{n+1} &= p^2(1-F) + pF + (1/2)(2pq)(1-F) \\ &= p^2(1-F) + pF + pq(1-F) \\ &= p^2 + pq + F(p - p^2 - pq) \\ &= p(p+q) + pF(1-p-q) \\ &= p(1) + pF(0) \\ &= p \end{aligned}$$

Thus, inbreeding does not change allelic frequencies. We can also see intuitively that inbreeding affects zygotic combinations (genotypes), but not allelic frequencies: Although inbreeding may determine the genotypes of offspring, inbreeding does not change the numbers of each allele that an individual transmits into the next generation.

In summary, inbreeding causes an increase in homozygosity, affects all loci in a population equally, and, in itself, has no effect on allelic frequencies, although it can expose deleterious alleles to selection. The results of inbreeding are evident in the appearance of recessive traits that are often deleterious. Inbreeding increases the rate of fetal deaths and congenital malformations in human beings and in other species that normally outbreed. In outbred agricultural crops and farm animals, decreases in size, fertility, vigor, and yield often result from inbreeding. Once deleterious traits appear due to inbreeding, natural selection can cause their removal from the population. However, in species adapted to inbreeding, including many crop plants and farm animals, inbreeding does not expose deleterious alleles because those alleles have generally been eliminated already.

Pedigree Analysis

Path diagram construction

The inbreeding coefficient, *F*, of an individual (the probability of autozygosity) can be determined by pedigree analysis. This is done by converting a pedigree to a *path diagram* by eliminating all extraneous individuals, those who cannot contribute to the inbreeding coefficient of the individual in question. A path diagram shows the direct line of descent from common ancestors. An example of the conversion of a pedigree to a path diagram is shown, in which individuals C and F are omitted from the path of descent because they are not related to anyone on the other side of the family tree and, therefore, do not contribute to the "common ancestry" of individual I. The pedigree shows an offspring who is the daughter of first cousins. Since first cousins are the offspring of siblings, they share a set of common grandparents. Thus, individual I can be autozygous for alleles from either ancestor A or B, her great-grandparents. The path diagram shows the only routes by which autozygosity can occur.

The inbreeding coefficient of the offspring of first cousins can be calculated as follows. Two paths of autozygosity appear in this diagram, one path for each grandparent as a common ancestor: A to D and E, then to G and H, and finally to I; or B to D and E, then to G and H, and finally to I.

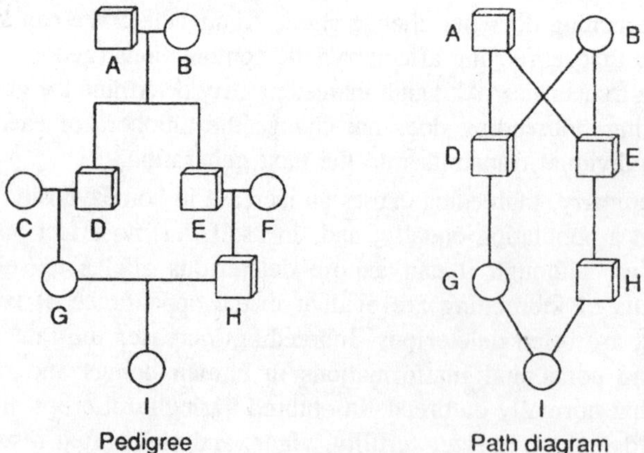

Fig. 4.1. Conversion of a pedigree to a path diagram.

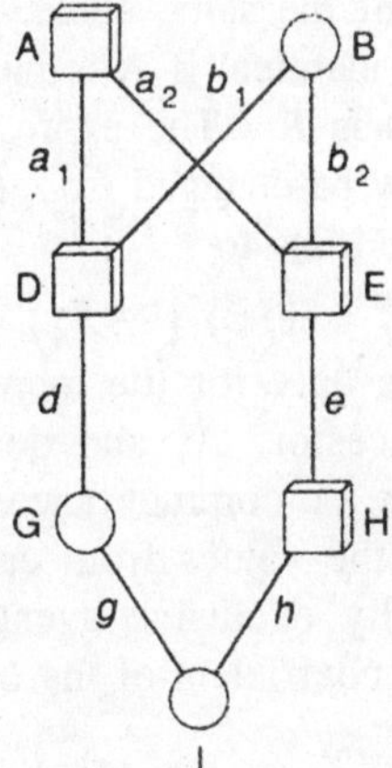

Fig. 4.2. The path diagram of the mating of first cousins with gametes labeled in lowercase letters.

In the path with A as the common ancestor, A contributes a gamete to D and a gamete to E. The probability is one-half that D and E each carry a copy of the same allele. That is, there are four possible allelic combinations for the two gametes, a_1 and a_2: *A-A*; *A-a*; *a-A*; and *a-a*. Of these combinations, the first and last (*A-A* and *a-a*) give a copy of the same allele to the two offspring, D and E, and can thus contribute to autozygosity. The probability that gametes a_1 and *d* carry copies of the same allele is one-half, and the probability that *d* and *g* carry copies of the same allele is also one-half. Similarly, on the other side of the pedigree, the probability is one-half that a_2 and *e* carry copies of the same allele and one-half that *e* and *h* carry copies of the same allele. Thus, the overall probability that the alleles that *g* and *h* carry are identical by descent (autozygous) is $(1/2)^5$. In general, it would be $(1/2)^n$ for each path, where *n* is the number of ancestors in the path.

You may have spotted an additional factor here. Of the possible combinations of allelic copies passed on to D and E, one-half (*A-A* and *a-a*) are autozygous combinations. However, the other half of the combinations, *A-a* and *a-A*, can lead to autozygosity if A is itself inbred. If we let F_A be the inbreeding coefficient of A (the probability that any two alleles at a locus in A are identical by descent), then F_A is the probability that the *A-a* and *a-A* combinations are also autozygous. Thus, the probability that a common ancestor, A, passes on copies of an identical ancestral allele is $1/2\ (1/2)F_A$, or $(1/2)\ (1 + F_A)$. In other words, there is a one-half probability that the alleles transmitted from

A to D and E are copies of the same allele. In the other half of the cases, these alleles can be identical if A is inbred. The probability of identity of A's two alleles is F_A. The expression for the inbreeding coefficient of I, F_I, can now be changed from $(1/2)^n$ by substituting $(1/2)(1 + F_A)$ for one of the (1/2)s to

$$F_I = (1/2)^n (1 - F_A)$$

This equation accounts only for the inbreeding of I by the path involving the common ancestor, A, and does not account for the symmetrical path with B as the common ancestor. To obtain the total probability of inbreeding, the values from each path must be added (because these are mutually exclusive events). Thus the complete formula for the inbreeding coefficient of the offspring of first cousins is

$$F_I = \Sigma[(1/2)^n (1 + F_J)] \qquad ...(1)$$

in which F_I is the probability that the two alleles in I are identical by descent, n is the number of ancestors in a given path, F_J is the inbreeding coefficient of the common ancestor of that path, and all paths are summed.

In the example of the mating of first cousins

$$F_I = (1/2)^5 (1 + F_A) + (1/2)^5 (1 + F_B)$$

If we assume that F_A and F_B are zero (which we must assume when the pedigrees of A and B are unknown), then

$$F_I = 2(1/2)^5 = (1/2)^4 = 0.0625$$

This can be interpreted to mean that about 6.25% of individual I's loci are autozygous, or that there is a 6.25% chance of autozygosity at any one of I's loci.

The inbreeding coefficient of the offspring of siblings can also be calculated, assuming that A and B are not themselves inbred (F_A and F_B are zero), as

$$F_I = 2(1/2)^3 = 0.25$$

Thus, about 25% of the loci in an offspring of siblings are autozygous.

Path diagram rules

The following points should be kept in mind when calculating an inbreeding coefficient:

1. All possible paths must be counted. A path is possible if gametes can actually pass in that direction. Paths that violate the rules of inheritance cannot be used. For example, the following path is unacceptable: I G E A D H I.

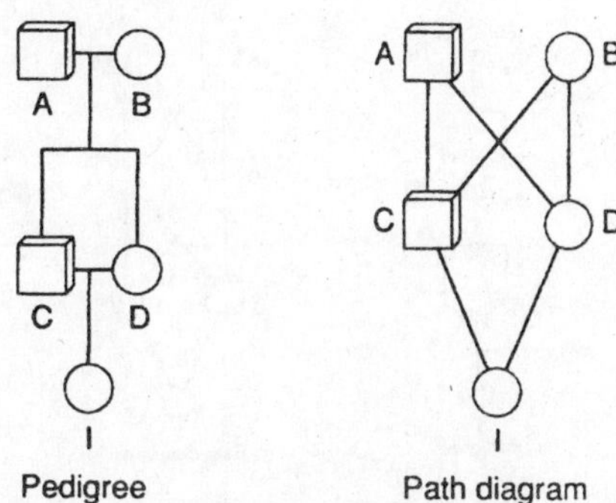

Fig. 4.3. Conversion of a sub-mating pedigree to a path diagram. Individual I is inbred.

2. In any path, an individual can be counted only once.
3. Every path must have one and only one common ancestor. The inbreeding coefficient of any other individual in the path is immaterial.

We present a complex pedigree produced from repeated sib mating, a pattern found in livestock and laboratory animals. This pedigree has two interesting points. First, common ancestors occur in several different generations. Second, some of the paths are complex. Thus, we must be sure to count all paths (paths 5 and 6 might not be immediately obvious). Although one of the common ancestors, A, is also inbred (F_A = 0.05)—a fact that we must take into consideration in paths 3 and 5. Thus, F_I is as follows:

From path 1: $(1/2)^3$	= 0.1250
From path 2: $(1/2)^3$	= 0.1250
From path 3: $(1/2)^5(1+0.05)$	= 0.0328
From path 4: $(1/2)^5$	= 0.0313
From path 5: $(1/2)^5(1 + 0.05)$	=0.0328
From path 6: $(1/2)^5$	= 0.0313
	F_I = 0.3782

Population Analysis

It is also possible to define the inbreeding coefficient, F, of a population as the relative reduction in heterozygosity in the population due to inbreeding. In an individual, F is the probability of autozygosity; it represents an increase in homozygosity, which is therefore a decrease in heterozygosity. In a population, it also represents the reduction in heterozygosity. From the definition, we can calculate the population F as follows:

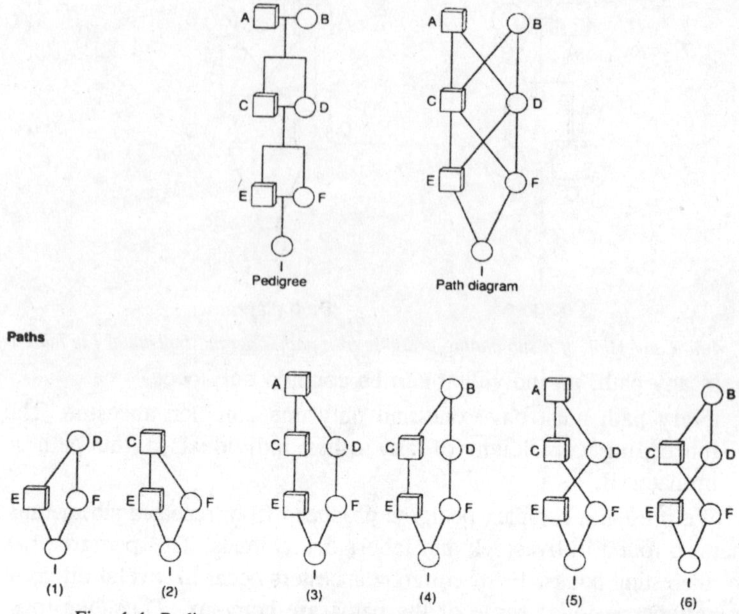

Pedigree

Path diagram

Paths

(1) (2) (3) (4) (5) (6)

Fig. 4.4. Pedigree and path diagram of two generations of sib matings.

$$F = \frac{(2pq - H)}{2pq}$$

where H is the actual proportion of heterozygotes in a population, and $2pq$ is the expected proportion of heterozygotes based on Hardy-Weinberg proportions. This equation reduces to

$$F = 1 - \frac{H}{2pq} \qquad ...(2)$$

This equation shows that when $H = 2pq$, F is zero, meaning that there is no decrease in heterozygotes and therefore, apparently, no inbreeding. When there are no heterozygotes, $F = 1$. This could be the case in a completely inbred population—for example, a self-fertilizing plant species.

As an example of an intermediate case, take the sample of one hundred individuals segregating the A_1 and A_2 alleles at the A locus: A_1A_1, fifty-four; A_1A_2, thirty-two; and A_2A_2, fourteen. In this example, $p = 0.7$, $q = 0.3$, and $H = 0.32$. Since $2pq = 0.42$, $H/2pq = 0.32/0.42 = 0.76$, and $F = 1 - 0.76$, or 0.24. Thus, the inbreeding coefficient of this population is 0.24; there is a 24% reduction in heterozygotes, due presumably to inbreeding.

5

POLYGENIC INHERITANCE

When we talked previously of genetic traits, we were usually discussing traits in which variation is controlled by single genes whose inheritance patterns led to simple ratios. However, many traits, including some of economic importance—such as yields of milk, corn, and beef—exhibit what is called *continuous* variation.

Although some variation occurred in height in Mendel's pea plants, all of them could be scored as either tall or dwarf; there was no overlap. Using the same methods that Mendel used, we can look at ear length in corn. With Mendel's peas, all of the F_1 were tall. In a cross between corn plants with long and short ears, all of the F_1 plants have ears intermediate in length between the parents. When both pea and corn F_1 plants are self-fertilized, the results are again different. In the F_2 generation, Mendel obtained exactly the same height categories (tall and dwarf) as in the parental generation. Only the ratio was different—3:1.

In corn, however, ears of every length, from the shortest to the longest, are found in the F_2; there are no discrete categories. A genetically controlled trait exhibiting this type of variation is usually controlled by many loci. In this chapter, we study this type of variation by looking at traits controlled by progressively more loci. We then turn to the concept of heritability, which is used as a statistical tool to evaluate the genetic control of traits determined by many loci.

TRAITS CONTROLLED BY MANY LOCI

Let us begin by considering grain colour in wheat. When a particular strain of wheat having red grain is crossed with another strain having white grain, all the F_1 plants have kernels intermediate

P_1	Red *AA*	×	White *aa*
F_1		Intermediate colour *Aa*	
F_2	Red : *AA*	Intermediate : *Aa*	White *aa*
		1 : 2 : 1	

Fig. 5.1. Cross involving the grain colour of wheat.

in colour. When these plants are self-fertilized, the ratio of kernels in the F_2 is 1 red: 2 intermediate: 1 white. This is inheritance involving one locus with two alleles. The white allele, *a,* produces no pigment (which results in the background colour, white); the red allele, *A*, produces red pigment. The F_1 heterozygote, *Aa,* is intermediate (incomplete dominance). When this monohybrid is self-fertilized, the typical 1:2:1 ratio results. (For simplicity, we use dominant-recessive allele designations, *A* and *a*. Keep in mind, however, that the heterozygote is intermediate in colour.)

Two-Locus Control

Now let us examine the same kind of cross using two other stocks of wheat with red and white kernels. Here, when the resulting intermediate (medium-red) F_1 are self-fertilized, five colour classes of kernels emerge in a ratio of 1 dark red:4 medium dark red:6 medium red:4 light red:1 white. The offspring ratio, in sixteenths, comes from the self-fertilization of a dihybrid in which the two loci are unlinked. In this case, both loci affect the same trait in the same way. In figure, each capital letter represents an allele that produces one unit of colour, and each lowercase letter represents an allele that produces no colour. Thus, the genotype *AaBb* has two units of colour, as do the genotypes *AAbb* and *aaBB*. All produce the same intermediate grain colour. Recall that a cross such as this produces nine genotypes in a

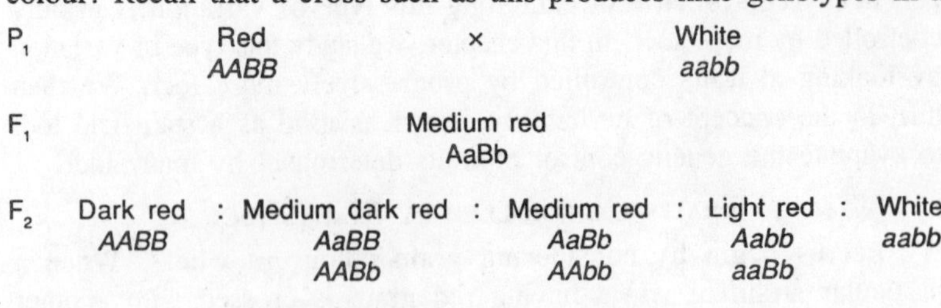

Fig. 5.2. Another cross involving grain colour of wheat.

ratio of 1:2:1:2:4:2:1:2:1. If these classes are grouped according to numbers of colour-producing alleles, the 1:4:6:4:1 ratio appears. This ratio is a product of a binomial expansion.

Three-Locus Control

In yet another cross of this nature, H. Nilsson-Ehle in 1909 crossed two wheat strains, one with red and the other with white grain, that yielded plants in the F_1 generation with grain of intermediate colour. When these plants were self-fertilized, at least seven colour classes, from red to white, were distinguishable in a ratio of 1:6:15:20:15:6:1. This result is explained by assuming that three loci are assorting independently, each with two alleles, so that one allele produces a unit of red colour and the other allele does not. We then see seven colour classes, from red to white, in the 1:6:15:20:15:6:1 ratio. This ratio is in sixty-fourths, directly from the 8 $\times$ 8 (trihybrid) Punnett square, and comes from grouping genotypes in accordance with the number of colour-producing alleles they contain. Again, the ratio is one that is generated in a binomial distribution.

Multilocus Control

From here, we need not go on to an example with four loci, then five, and so on. We have enough information to draw generalities. It should not be hard to see how discrete loci can generate a continuous distribution. Theoretically, it should be possible to distinguish different colour classes down to the level of the eye's ability to perceive differences in wavelengths of light. In fact, we rapidly lose the ability to assign unique colour classes to genotypes because the variation within each genotype soon causes the phenotypes to overlap. For example, with three loci, a colour somewhat lighter than medium dark red may belong to the medium-dark-red class with three colour alleles, or it may belong to the medium-red class with only two colour alleles.

The variation within each genotype is due to the environment—that is, two organisms with the same genotype may not necessarily be identical in colour because nutrition, physiological state, and many other variables influence the phenotype. It is possible for the environment to obscure genotypes even in a one-locus, two-allele system. That is, a height of 17 cm could result in the F_2 from either the *aa* or *Aa* genotype in the figure when there is excessive variation. In the other two cases, there would be virtually no organisms 17 cm tall. Systems such as those we are considering, in which each allele contributes a small unit to the phenotype, are easily influenced by the environment,

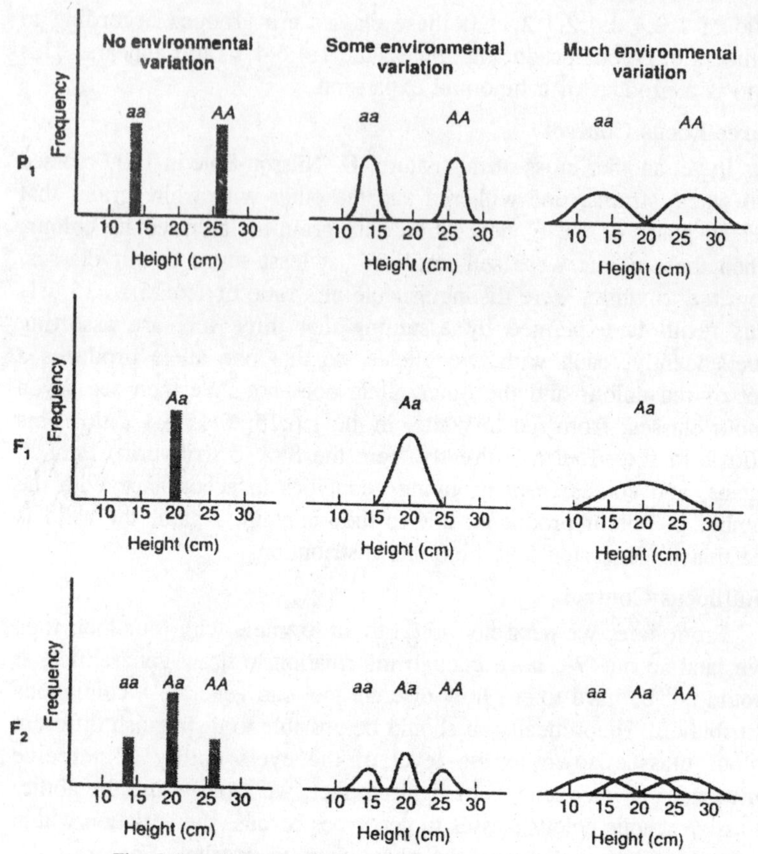

Fig. 5.3. Influence of environment on phenotype distribution.

with the result that the distribution of phenotypes approaches the bell-shaped curve seen at the bottom.

Thus, phenotypes determined by multiple loci with alleles that contribute dosages to the phenotype will approach a continuous distribution. This type of trait is said to exhibit *continuous*, *quantitative*, or *metrical variation*. The inheritance pattern is *polygenic* or *quantitative*. The system is termed an *additive model* because each allele adds a certain amount to the phenotype.

From the three wheat examples just discussed, we can generalize to systems with more than three polygenic loci, each segregating two alleles. We can predict the distribution of genotypes and phenotypes expected from an additive model with any number of unlinked loci segregating two alleles each. This table is useful when we seek to estimate how many loci are producing a quantitative trait, assuming it is possible to distinguish the various phenotypic classes. For example, when a strain of heavy mice was crossed with a lighter strain, the F_1 were of intermediate weight. When these F_1 were interbred, a continuous distribution of adult weights appeared in the F_2 generation. Since only about one mouse in 250 was as heavy as the heavy parent stock, we could guess that if an additive model holds, then four loci are segregating. This is because we expect $1/(4)^n$ to be as extreme as either parent; one in 250 is roughly $1/(4)^4 = 1/256$.

Location of Polygenes

The fact that traits with continuous variation can be controlled by genes dispersed over the whole genome was shown by James Crow, who studied DDT resistance in *Drosophila*. A DDT-resistant strain of flies was created by growing them on increasing concentrations of the insecticide. Crow then systematically tested each chromosome for the amount of resistance it conferred. Susceptible flies were mated with resistant flies, and the sons from this cross were backcrossed. Offspring were then scored for the particular resistant chromosomes they contained (each chromosome had a visible marker) and were tested for their resistance to DDT. Sons were used in the backcross because there is no crossing over in males. Therefore, the sons would pass resistant and susceptible chromosomes on intact. As you can see, each chromosome has the potential to increase the fly's resistance to DDT. Thus, each chromosome contains loci (polygenes) that contribute to the phenotype of this additive trait.

Significance of Polygenic Inheritance

The concept of additive traits is of great importance to genetic theory because it demonstrates that Mendelian rules of inheritance can explain traits that have a continuous distribution—that is, Mendel's rules for discrete characteristics also hold for quantitative traits. Additive traits are also of practical interest. Many agricultural products, both plant and animal, exhibit polygenic inheritance, including milk production and fruit and vegetable yield. In addition, many human traits, such as height and IQ, appear to be polygenic, although with substantial environmental components.

Historically, the study of quantitative traits began before the rediscovery of Mendel's work at the turn of the century. In fact, biologists in the early part of this century debated as to whether the "Mendelians" were correct or whether the "biometricians" were correct in regard to the rules of inheritance. Biometricians used statistical techniques to study traits characterized by continuous variation and claimed that single discrete genes were not responsible for the observed inheritance patterns. They were interested in evolutionarily important facets of the phenotype—traits that can change slowly over time. Mendelians claimed that the phenotype was controlled by discrete "genes." Eventually the Mendelians were proven correct, but the biometricians' tools were the only ones suitable for studying quantitative traits.

The biometric school was founded by F. Galton and K. Pearson, who showed that many quantitative traits, such as height, were inherited. They invented the statistical tools of correlation and regression analysis in order to study the inheritance of traits that fall into smooth distributions.

Population Statistics

A distribution can be described in several ways. One is the formula for the shape of the curve formed by the frequencies within the distribution. A more functional description of a distribution starts by defining its center, or *mean*. As we can see from the figure, the mean is not itself enough to describe the distribution. Variation about this mean determines the actual shape of the curve. (We confine our discussion to symmetrical, bell-shaped curves called *normal distributions*. Many distributions approach a normal distribution.)

Mean, Variance, and Standard Deviation

The mean of a set of numbers is the arithmetic average of the numbers and is defined as

$$\bar{x} = \Sigma x / n \qquad \text{...(1)}$$

in which

$\bar{x}$ = the mean

Σx = the summation of all values

n = the number of values summed

The variation about the mean is calculated as the average squared deviation from the mean:

$$s^2 = V = \frac{\Sigma(x - \bar{x})^2}{n - 1} \qquad \text{...(2)}$$

This value (V or s^2) is called the *variance*. Observe that the flatter the distribution is, the greater the variance will be.

The variance is one of the simplest measures we can calculate of variation about the mean. You might wonder why we simply don't

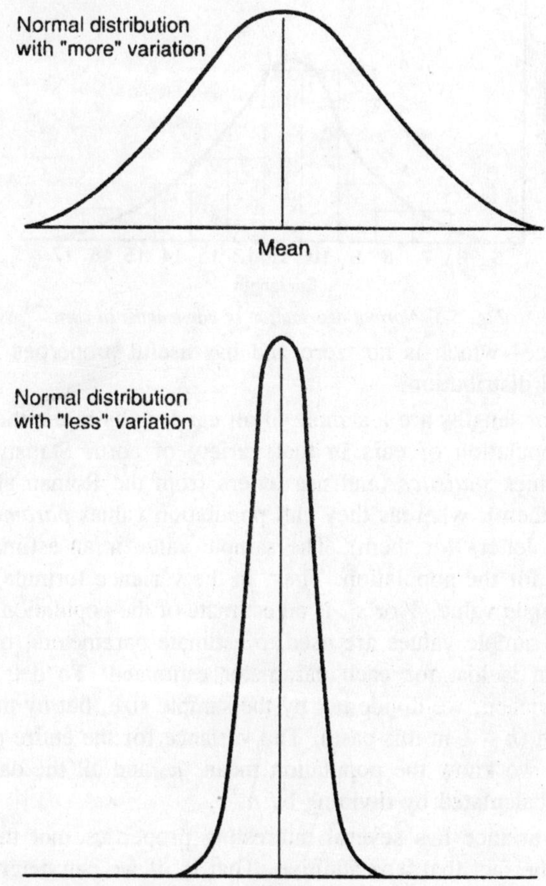

Fig. 5.4. Two normal distributions (bell-shaped curves) with the same mean.

calculate an average deviation from the mean rather than an average squared deviation. For example, we could calculate a measure of variation as

$$\frac{\Sigma(x - \overline{x})}{n - 1}$$

(We will get to why we use $n - 1$ rather than n in the denominator in a moment.) Note, however, that the above measure is zero. By the definition of the mean, the absolute value of the sum of deviations above it is equal to the absolute value of the sum of deviations below it—one is negative and the other is positive. However, by squaring each deviation, as in equation 2, we create a relatively simple index—

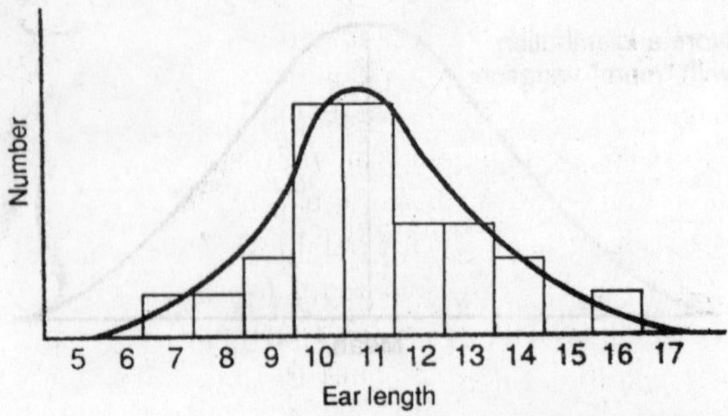

Fig. 5.5. Normal distribution of ear lengths in corn.

the variance—which is not zero and has useful properties related to the normal distribution.

The ear lengths are a sample of all ear lengths in the theoretically infinite population of ears in that variety of corn. Statisticians call sample values *statistics* (and use letters from the Roman alphabet to represent them), whereas they call population values *parameters* (and use Greek letters for them). The sample value is an estimate of the true value for the population. Thus, in the variance formula (equation 2), the sample value, V or s^2, is an estimate of the population variance, σ^2. When sample values are used to estimate parameters, one degree of freedom is lost for each parameter estimated. To determine the sample variance, we divide not by the sample size, but by the degrees of freedom ($n - 1$ in this case). The variance for the entire population (assuming we know the population mean, μ, and all the data values) would be calculated by dividing by n.

The variance has several interesting properties, not the least of which is the fact that it is additive. That is, if we can determine how much a given variable contributes to the total variance, we can subtract that amount of variance from the total, and the remainder is caused by whatever other variables (and their interactions) affect the trait. This property makes the variance extremely important in quantitative genetic theory.

The *standard deviation* is also a measure of variation of a distribution. It is the square root of the variance:

$$s = \sqrt{V} \qquad \ldots(3)$$

In a normal distribution, approximately 67% of the area of the curve lies within one standard deviation on either side of the mean, 96% lies within two standard deviations, and 99% lies within three standard deviations. Thus, about two-thirds of the population would have ear

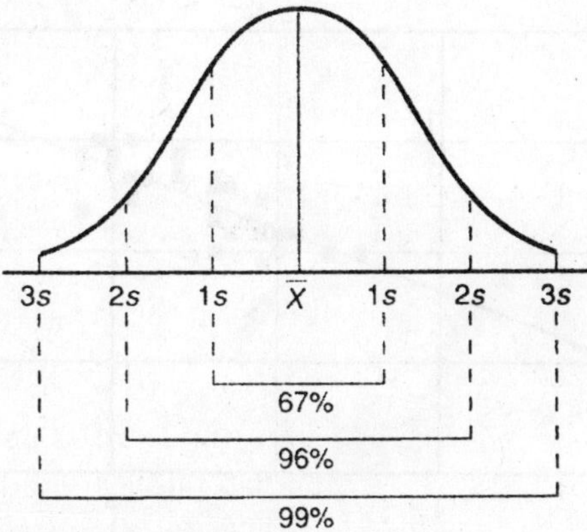

Fig. 5.6. Area under the bell-shaped surve.

lengths between 9.12 and 13.12 cm (mean ± standard deviation). One final measure of variation about the mean is the *standard error of the mean* (SE):

$$SE = s\sqrt{n}$$

The standard error (of the mean) is the standard deviation about the mean of a distribution of sample means. In other words, if we repeated the experiment many times, each time we would generate a mean value. We could then use these mean values as our data points. We would expect the variation among a population of means to be less than among individual values, and it is. Data are often summarized as "the mean ± SE." In our example, SE = $2.0/\sqrt{25}$ = 2.0/5.0 = 0.4. We can summarize the data set as 11.1 ± 0.4 (mean ± SE).

Covariance, Correlation, and Regression

It is often desirable in genetic studies to know whether a relationship exists between two given characteristics in a series of individuals. For example, is there a relationship between height of a plant and its weight, or between scholastic aptitude and grades, or between a phenotypic measure in parents and their offspring? If one increases, does the other also? With increasing wing length in mid-parent (the average of the two parents: x-axis), there is an increase in offspring wing length (y-axis). We can determine how closely the two variables are related by calculating a *correlation coefficient*—an index that goes from –1.0 to + 1.0, depending on the degree of relationship between the variables. If there is no relation (if the variables are independent), then the correlation coefficient will be zero. If there is perfect correlation, where an increase in one variable is associated

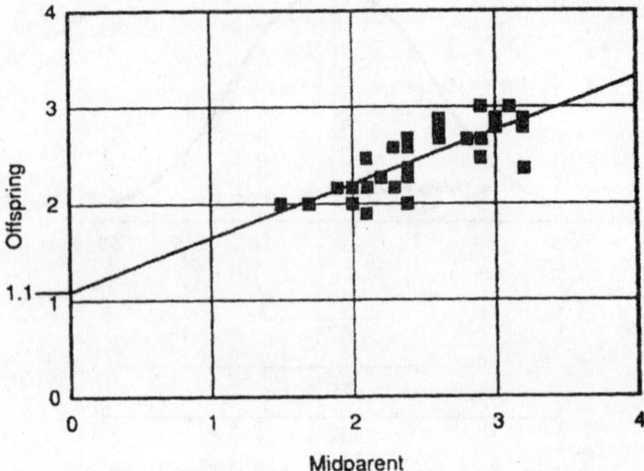

Fig. 5.7. The relationship between two variables, parental and offspring wing length in fruit flies.

with a proportional increase in the other, the coefficient will be + 1.0. If an increase in one is associated with a proportional decrease in the other, the coefficient will be –1.0. The formula for the correlation coefficient (r) is

$$r = \frac{\text{convariance of } x \text{ and } y}{s_x \cdot s_y} \quad ...(4)$$

where s_x and s_y are the standard deviations of x and y, respectively.

To calculate the correlation coefficient, we need to define and calculate the *covariance* of the two variables, cov(x, y). The covariance is analogous to the variance, but it involves the simultaneous deviations from the means of both the x and y variables:

$$\text{cov}(x,y) = \frac{\Sigma(x-\overline{x})(y-\overline{y})}{n-1} \quad ...(5)$$

The analogy between variance and covariance can be seen by comparing equations 5 and 2. The variances, standard deviations, and covariance are calculated, in which the correlation coefficient, r, is 0.78. (There are computational formulas available that substantially cut down on the difficulty of calculating these statistics. If a computer or calculator is used, only the individual data points need to be entered—most computers and many calculators can be programmed to do all the computations.)

Many experiments deal with a situation in which we assume that one variable is dependent on the other (in a cause-and-effect relationship). For example, we may ask, what is the relationship of

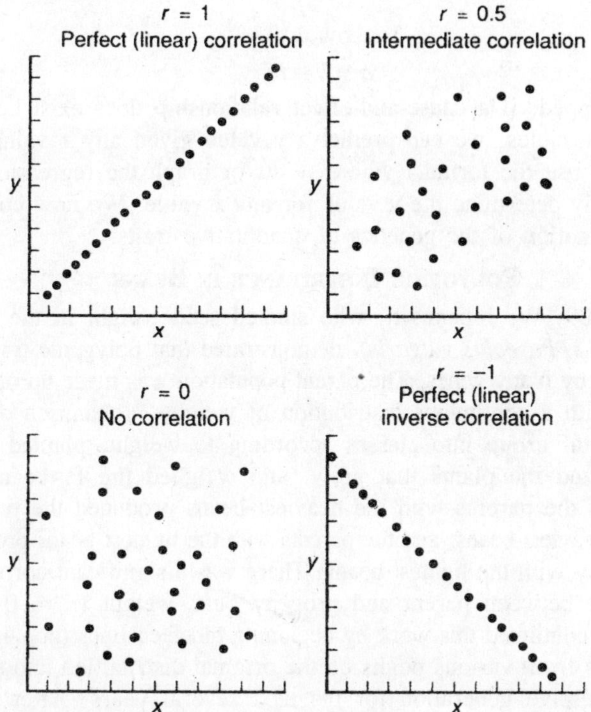

Fig. 5.8. Plots showing varying degrees of correlation within data sets.

DDT resistance in *Drosophila* to an increased number of DDT-resistant alleles? With more of these alleles, the DDT resistance of the flies should increase. Number of DDT-resistant alleles is the independent variable, and resistance of the flies is the dependent variable. That is, a fly's resistance is dependent on the number of DDT-resistant alleles it has, not the other way around. Going back, we could make the assumption that offspring wing length is dependent on parental wing length. If this were so, a technique called *regression analysis* could be used. This analysis allows us to predict an offspring's wing length (y variable) given a particular mid-parental wing length (x variable). (It is important to note that regression analysis assumes a cause-and-effect relationship, whereas correlation analysis does not.)

The formula for the straight-line relationship (regression line) between the two variables is $y = a + bx$, where b is the slope of the line (change in y divided by change in x, or $\Delta y/\Delta x$) and a is the y-intercept of the line. To define any line, we need only to calculate the slope, b, and the y intercept, a:

$$b = \text{cov}(x, y) / s_x^2 \qquad ...(6)$$

$$a = \bar{y} - b\bar{x} \qquad ...(7)$$

Thus equipped, if a cause-and-effect relationship does exist between the two variables, we can predict a *y* value given any *x* value. We can either use the formula $y = a + bx$ or graph the regression line and directly determine the *y* value for any *x* value. We now continue our examination of the genetics of quantitative traits.

Polygenic Inheritance in Beans

In 1909, W. Johannsen, who studied seed weight in the dwarf bean plant (*Phaseolus vulgaris*), demonstrated that polygenic traits are controlled by many genes. The parent population was made up of seeds (beans) with a continuous distribution of weights. Johannsen divided this parental group into classes according to weight, planted them, self-fertilized the plants that grew, and weighed the F_1 beans. He found that the parents with the heaviest beans produced the progeny with the heaviest beans, and the parents with the lightest beans produced the progeny with the lightest beans. There was a significant correlation coefficient between parent and progeny bean weight ($r = 0.34 \pm 0.01$). He continued this work by beginning nineteen lines (populations) with beans from various points on the original distribution and selfing each successive generation for the next several years. After a few generations, the means and variances stabilized within each line. That is, when Johannsen chose, within each line, parent plants with heavier-than-average or lighter-than-average seeds, the offspring had the parental mean with the parental variance for seed size. For example, in one line, plants with both the lightest average bean weights (24 centigrams) and plants with the heaviest average bean weights (47 cg) produced offspring with average bean weights of 37 cg. By selfing the plants each generation, Johannsen had made them more and more homozygous, thus lowering the number of segregating polygenes. Therefore, the lines became homozygous for certain of the polygenes, and any variation in bean weight was then caused only by the environment. Johannsen thus showed that quantitative traits were under the control of many segregating loci.

Selection Experiments

Selection experiments are done for several reasons. Plant and animal breeders select the most desirable individuals as parents in order to improve their stock. Population geneticists select specific characteristics for study in order to understand the nature of quantitative genetic control. For example, *Drosophila* were tested in a fifteen-choice maze for geotactic response.

The maze was on its side, so at every intersection, a fly had to make a choice between going up or going down. The flies with the highest scores were chosen as parents for the "high" line (positive geotaxis; favoured downward direction), and the flies with the lowest score were chosen as parents for the "low" line (negative geotaxis; favoured upward direction). The same selection was made for each generation. As time progressed, the two lines diverged quite significantly. This tells us that there is a large genetic component to the response; the experimenters are successfully amassing more of the "downward" alleles in the high line and more of the "upward" alleles in the low line. Several other points emerge from this graph. First, the high and low responses are slightly different, or asymmetrical. The high line responded more quickly, leveled out more quickly, and tended toward the original state more slowly after selection was relaxed. (The relaxation of selection occurred when the parents were a random sample of the adults rather than the extremes for geotactic scores.) The low line responded more slowly and erratically. In addition, the low line returned toward the original state more quickly when selection was relaxed.

The nature of these responses indicates that the high line became more homozygous than the low line. This is shown by the former's response when selection is relaxed: It has exhausted a good deal of its variability for the polygenes responsible for geotaxis. The low line, however, seems to have much of its original genetic variability, because the relaxation of selection caused the mean score of this line to increase rapidly. It still had enough genetic variability to head back to the original population mean. The response to a selection experiment is one way that plant and animal breeders can predict future response.

HERITABILITY

Plant and animal breeders want to improve the yields of their crops to the greatest degree they can. They must choose the parents of the next generation on the basis of this generation's yields; thus, they are continually performing selection experiments. Breeders run into two economic problems. They cannot pick only the very best to be the next generation's parents because (1) they cannot afford to decrease the size of a crop by using only a very few select parents and (2) they must avoid *inbreeding depression*, which occurs when plants are self-fertilized or animals are bred with close relatives for many generations. After frequent inbreeding, too much homozygosity occurs, and many genes that are slightly or partially deleterious begin to show themselves, depressing vigor and yield. Thus, breeders need some index of the potential response to selection so that they can then get the greatest amount of selection with the lowest risk of inbreeding depression.

Realized Heritability

Breeders often calculate a *heritability* estimate, a value that predicts to what extent their selection will be successful. Heritability is defined in the following equation:

$$H = \frac{Y_O - \bar{Y}}{Y_P - \bar{Y}} = \frac{\text{gain}}{\text{selection differential}} \quad \text{...(8)}$$

in which

H = heritability

Y_O = offspring yield

$\bar{Y}$ = mean yield of the population

Y_P = parental yield

From this equation, we can see that heritability is the gain in yield divided by the amount of selection practiced. $Y_O - \bar{Y}$ is the improvement

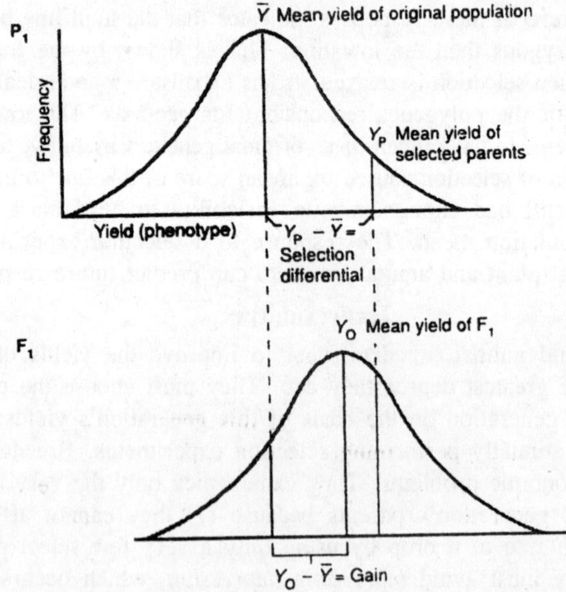

Fig. 5.9. Realized heritability is the gain divided by the selection differential.

over the population average due to $Y_P - \bar{Y}$, which is the amount of difference between the parents and the population average. If there is no gain ($Y_O = \bar{Y}$), then the heritability will be zero, and breeders will know that no matter how much selection they practice, they will not improve their crops and might as well not waste their time. Since this value is calculated after the breeding has been done, it is referred to as *realized heritability*.

The following example may help to clarify the calculation of realized heritability. The number of bristles on the sternopleurite, a thoracic plate in *Drosophila*, is under polygenic control. In a population of flies, the mean bristle number was 6.4. Three pairs of flies served as parents; they had a mean of 7.2 bristles. Their offspring had a mean of 6.6 bristles. Hence, $Y_O = 6.6$, $\bar{Y} = 6.4$, and $Y_P = 7.2$. Dividing the gain by the selection differential—that is, substituting in equation 8—gives us

$$H = \frac{6.6 - 6.4}{7.2 - 6.4} = \frac{0.2}{0.8} = 0.25$$

If both a low line and a high line were begun, and if both were carried over several generations, the heritability would be measured by the final difference in means of the high and low lines (gain) divided by the cumulative selection differentials summed for both the high and low lines.

Note that the response to selection declines with time as the selected population becomes homozygous for various alleles controlling the trait. As the response declines, the calculated heritability value itself declines. After intense or prolonged selection, heritability may be zero. It does not mean that the trait is not controlled by genes, only that there is no longer a response to selection. Hence, heritability is specific for a particular population at a particular time. Intense selection exhausts the genetic variability, rendering the response to selection, and thus the heritability itself, zero.

Quantitative geneticists treat the realized heritability as an estimate of *true heritability*. True heritability is actually viewed in two different ways: as *heritability in the narrow sense* and *heritability in the broad sense*. We define these on the basis of partitioning of the variance of the quantitative character under study.

Partitioning of the Variance

Given that the variance of a distribution has genetic and environmental causes, and given that the variance is additive, we can construct the following formula:

$$V_{Ph} = V_G + V_E \qquad ...(9)$$

in which

V_{Ph} = total phenotypic variance

V_G = variance due to genotype

V_E = variance due to environment

Throughout the rest of this discussion, we will stay with this model. We could construct a more complex variance model if there are interactions between variables. For example, if one genotype responded better in one soil condition than in another soil condition, this environment-genotype interaction would require a separate variance term (V_{GE}).

The variance due to the genotype (V_G) can be further broken down according to the effects of additive polygenes (V_A), dominance (V_D), and epistasis (V_I) to give us a final formula:

$$V_{Ph} = V_A + V_D + V_I + V_E \qquad ...(10)$$

We can now define the two commonly used—and often confused—measures of heritability. *Heritability in the narrow sense* is

$$H_N = V_A/V_{Ph} \qquad ...(11)$$

This heritability is the proportion of the total phenotypic variance caused by additive genetic effects. It is the heritability of most interest to plant and animal breeders because it predicts the magnitude of the response under selection.

Heritability in the broad sense is

$$H_B = V_G/V_{Ph} \qquad ...(12)$$

This heritability is the proportion of the total phenotypic variance caused by all genetic factors, not just additive factors. It measures the extent to which individual differences in a population are caused by genetic differences. This measure is the one most often used by psychologists. We are concerned primarily with H_N, heritability in the narrow sense.

Measurement of Heritability

Three general methods are used to estimate heritability. First, as discussed earlier, we can measure heritability by the response of a population to selection. Second, we can directly estimate the components of variance by minimizing one component; the remaining variance can then be attributed to other causes. For example, by minimizing environmental causes of variance, we can estimate the genetic component directly. Or, by eliminating the genetic causes of variance, we can estimate the environmental component directly. Third, we can measure the similarity between relatives. Variance components can be minimized in several different ways. If we use genetically identical organisms, then the additive, dominance, and epistatic variances are zero, and all that is left is the environmental variance. For example, F. Robertson determined the variance components for the length of the thorax in *Drosophila*. The total variance (V_{Ph}) in a genetically heterogeneous population was 0.366. He then looked at the variance in flies that were genetically homogeneous. These were from isolated lines inbred in the laboratory over many generations to become virtually homozygous. Robertson studied the F_1 in several different matings of inbred lines and found the variance in thorax length to be 0.186 (V_E). By subtraction (0.366 – 0.186), we know that the total genetic variance (V_G) was 0.180. From this, we can calculate heritability in the broad sense as

$$H_B = V_G/V_{Ph} = 0.180/0.366 = 0.49$$

To calculate a heritability in the narrow sense, it is necessary to extract the components of the genetic variance, V_G.

Genetic variance can be measured directly by minimizing the influence of the environment. This is most easily done with plants grown in a greenhouse. Under that circumstance, environmental variables, such as soil quality, water, and sunlight, can be controlled to a very high degree. Hence, the variance among individuals grown under these circumstances is almost all genetic variance. The total phenotypic variance can be obtained from the plants grown under natural circumstances. This allows us to calculate heritability in the broad sense.

Several methods exist to sort out the additive from the dominant and epistatic portions of the genetic variance. The methods rely mostly on correlations between relatives. That is, the expected amount of genetic similarity between certain relatives can be compared with the actual similarity. The expected amount of genetic similarity is the proportion of genes shared; this is a known quantity for any form of relatedness. For example, parents and offspring have half their genes in common. The relation of observed and expected correlations between relatives is a direct measure of heritability in the narrow sense. We can thus define

$$H_N = r_{obs}/r_{exp} \qquad ...(13)$$

in which r_{obs} is the observed correlation between the relatives, and r_{exp} is the expected correlation. The expected correlation is simply the proportion of the genes in common.

We must point out that the observed correlation between relatives can be artificially inflated if the environments are not random. Since we know that relatives frequently share similar (or correlated) environments, they may show a phenotypic similarity irrespective of genetic causes. It is important to keep that in mind, especially when we analyze human traits, where it may be almost impossible to rule out or quantify environmental similarity. Hence, r_{obs} may be inflated, which will inflate H_N.

In human beings, finger-ridge counts (fingerprints) have a very high heritability; there seems to be very little environmental interference in the embryonic development of the ridges. Monozygotic twins are from the same egg, which divides into two embryos at a very early stage. They have identical genotypes. Dizygotic twins result from the simultaneous fertilization of two eggs. They have the same genetic relationship as siblings. (However, environmental influences may be different; they may be treated differently by relatives and friends.) The data therefore suggest that human finger ridges are almost completely controlled by additive genes with a negligible input from environmental and dominance variation. Few human traits are controlled this simply.

This brief discussion should make it clear that the components of the total variance can be estimated. For a given quantitative trait, the total variance can be measured directly. If identical genotypes can be used, then the environmental component of variance can be determined. By correlation of various relatives, it is possible to directly measure heritability in the narrow sense. If heritability is known, and if the total phenotypic variance is known, then all that are left, assuming no interaction, are the dominance and epistatic components. In practice, the epistatic components are usually ignored. Thus, operationally, all that is left is the dominance variance, obtained by subtracting the additive from the total genetic variance. In addition, plant and animal breeders use sophisticated statistical techniques of covariance and variance analysis, techniques that are beyond our scope.

Quantitative Inheritance in Human Beings

As with most human studies, the measurement of heritability is limited by a lack of certain types of information. We cannot develop pure human lines, nor can we manipulate human beings into various kinds of environments or do selection experiments. However, certain kinds of information are available that allow some estimation of heritabilities.

Skin Colour

Skin colour is a quantitative human trait for which a simple analysis can be done on naturally occurring matings. Certain groups of people have black skin; other groups do not. Many of these groups breed true in the sense that skin colours stay the same generation after generation within a group; when groups intermarry and produce offspring, the F_1 are intermediate in skin colour. In turn, when F_1 individuals intermarry and produce offspring, the skin colour of the F_2 is, on the average, about the same as the F_1, but with more variation. The data are consistent with a model of four loci, each segregating two alleles. At each locus, one allele adds a measure of colour, whereas the other adds none.

IQ and Other Traits

In human beings, twin studies have been helpful in estimating the heritability of quantitative traits. One way of looking at quantitative traits is by the *concordance* among twins. Concordance means that if one twin has the trait, the other does also. Discordance means one has the trait and the other does not. High concordance of monozygotic as compared with dizygotic twins is another indicator of the heritability of a trait. Concordance values for measles susceptibility and handedness, which are similar for both monozygotic and dizygotic twins, demonstrate the environmental influence on some traits.

Some monozygotic twins (MZ) have been reared apart. The same is true for dizygotic twins (DZ) and nontwin siblings. IQ (intelligence quotient) is a measure of intelligence highly correlated among relatives, indicating a strong genetic component. In three studies of monozygotic twins reared apart, the average correlation in IQ was 0.72. In thirty-four studies of monozygotic twins reared together, the average correlation in IQ was 0.86; dizygotic twins reared together have an average correlation of 0.60 in IQ. Thus, it is clear that there is a genetic influence on IQ. However, experts disagree strongly on the environmental role in shaping IQ and the exact meaning of IQ as a functional measure of intelligence.

At present, twin studies are emerging from the shadow of a scandal involving a knighted British psychologist, Cyril Burt (1883–1971), who did classical twin research on the inheritance of IQ. Burt was posthumously accused of fraud, an accusation that was almost universally accepted and that cast doubt on all of his data and conclusions. More recently, new information seemed to cast doubt on the charges of fraud. These on-again, off-again charges have been a focus of scientific interest.

6

NONRANDOM MATING

Most natural populations deviate in some way from the random mating ideal considered thus far. For example, a species whose range exceeds the distance an individual moves in its lifetime cannot possibly mate at random. Departures from random mating can have profound consequences on the evolutionary dynamics of a species. The various ways in which a species might deviate from random mating are diverse, but we will concentrate on only two important departures from random mating, inbreeding and population subdivision. Both may be examined by first describing a generalized form of the Hardy-Weinberg law that includes a new state variable, F, and then by showing the dependency of F on the level of inbreeding or subdivision.

Inbreeding occurs when individuals are more likely to mate with relatives than with randomly chosen individuals. Inbreeding increases the probability that offspring are homozygous and, if practiced by a non-zero fraction of the population, that individuals in the population are homozygous. You will recall that the heterozygous effects of alleles affecting fitness traits are generally less than one-half; thus, this increase in the frequency of homozygotes has a detrimental effect. Inbreeding depression is the phenomenon of lower fitness with higher levels of inbreeding. This phenomenon is quite general: the fitness of most species decreases with increasing homozygosity. For some, the decrease is dramatic, with complete infertility or inviability after only a few generations of brother-sister mating. Inbreeding depression in humans will be described in this chapter, followed by a discussion of the role of inbreeding depression in the evolution of selfing in plants. A species with restricted migration will appear to be inbred because there are more homozygotes than expected under the assumption of random mating, a condition known as Wahlund's effect. Sewall Wright invented a set of measures of departures from Hardy-Weinberg for subdivided populations called F statistics. Here we will consider only the simplest of these, which is called F_{ST}, and show how F_{ST} has been used to investigate the effects of migration on geographic variation in a species.

GENERALIZED HARDY-WEINBERG

An assumption of the Hardy-Weinberg law is violated when populations do not mate at random. We write the genotype frequencies in this case as follows:

Genotype:	A_1A_1	A_1A_2	A_2A_2
Frequency:	$p^2(1 - F) + pF$	$2pq(1 - F)$	$q^2(1 - F) + qF$

The new state variable, F, is used to describe the deviation of the genotype frequencies from the Hardy-Weinberg frequencies. When $F = 0$, we recover the usual Hardy-Weinberg frequencies. If $0 < F \leq 1$, there is an excess of homozygotes compared to the Hardy-Weinberg expectation. When $F < 0$, there is an excess of heterozygotes.

Note that the pair of state variables p and F are always sufficient to describe the genotype frequencies at a single locus with two alleles. At first, this may seem incorrect because two variables should not be able to describe fully the state of a three-variable system. However, although there are three genotype frequencies, x_{11}, x_{12}, and x_{22}, they must add to one. Thus, the dimensionality of the space of genotype frequencies is two rather than three.

F has played such an important role in population genetics that it has acquired a myriad of interpretations. Most of these occur when F is between zero and one, as occurs with inbreeding and subdivision. In this instance, F may be interpreted as the probability of homozygosity due to special circumstances, PHSC (pronounced phonetically). For example, the frequency of A_1A_1 individuals in the population or, equivalently, the probability that a randomly chosen individual is A_1A_1 may be obtained by the following argument.

Draw an individual at random from the population. The probability that one of its two alleles at the A locus is A_1 is p. The probability that the other allele is also A_1, given that the first is A_1, is $F + (1 - F)p$. This latter event includes two mutually exclusive components: either the second allele is A_1 because of homozygosity due to special circumstances, which occurs with probability F, or it is not, which occurs with probability $(1 - F)$. If the individual is not homozygous due to special circumstances, the second allele may still be an A_1 allele, as the second allele is, in effect, drawn at random from the population; thus, the probability that it is an A_1 allele is p. We have established that the probability that the second allele is A_1 given that the first is A_1, is $F + (1 - F)p$. The joint probability of A_1 for the first and second alleles in the chosen individual is now seen to be $p[F + (1 - F)p]$, *as* given above.

A second interpretation of F, when it is between zero and one, is as the correlation of uniting gametes.

Now that we have a general way of describing the state of a population, the next step is to explore how nonrandom mating affects F. *We* first consider inbreeding, which occurs when relatives are preferentially chosen as mates. The quantitative aspects of inbreeding depend on the degree of relatedness of mates, which will be explored in the next section.

Identity by Descent

There are many occasions where a quantitative measure of the relationship between relatives is required. In this chapter, we need such a measure to describe inbreeding. In the next chapter, we need it to find the correlation between relatives for quantitative traits. Many different measures are possible; ours, naturally, will be based on the genetic relatedness of relatives. Relatives are genetically similar because they share alleles that are descended from common ancestor alleles. Recall that two alleles at the same locus that are descended from the same ancestral allele somewhere in their recent pasts are said to be identical by descent. Thus, identity by descent is a natural concept to use as the basis of a quantitative description of the relatedness of relatives. One simple measure of relatedness that uses identity by descent is the coefficient of kinship and is usually notated as f_{xy}. The coefficient of kinship is the probability that two alleles, one from individual X and one from individual Y, are identical by descent.

The coefficient of kinship is easy to calculate for simple pedigrees. For complicated pedigrees, one usually uses one of several available algorithms. As all of the pedigrees used in this book and most of population genetics are simple and as the act of reasoning through a pedigree is itself instructive, we will argue from first principles.

The simplest pair of relatives is a parent and its offspring, as illustrated here. In the figure, the offspring is labeled X and the parent, Y. To find the coefficient of kinship in this case, begin by choosing an

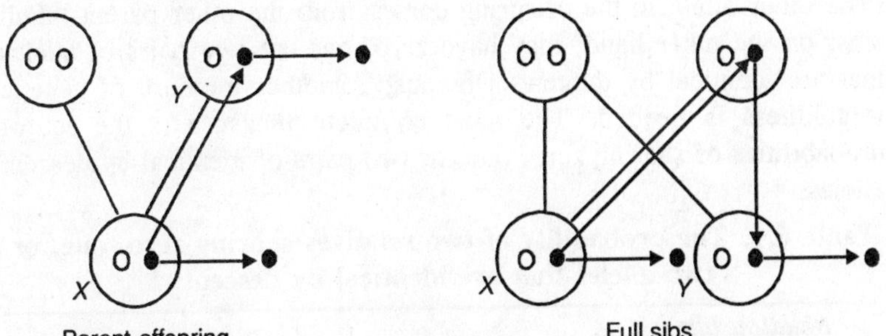

Fig. 6.1. Two pedigrees used in the text to illustrate the calculation of the coefficient of kinship.

allele at random from the offspring and an allele at random from the parent, as illustrated in the figure by the arrows pointing away from alleles in X and Y. The probability that these two alleles are identical by descent is 1/4, which is obtained in two steps. First, the probability that the allele chosen from X came from Y is 1/2. (It is equally likely that this allele came from either parent.) Second, given that the allele came from Y, the probability that it is the one originally chosen from Y is also 1/2. As these two events are independent, the total probability that the two chosen alleles are identical by descent is

$$f_{OP} = \frac{1}{2} \times \frac{1}{2} = \frac{1}{4}.$$

Here one sibling is X and the other is Y. As before, trace backward from the allele chosen from X. This time, the allele will be descended from one of the four alleles in the two parents with certainty. The probability that a copy of the ancestor allele ends up in individual Y is one-half; the probability that that same allele is the chosen allele is also one-half. Thus, $f_{FS} = 1/4$, just as for parent and offspring.

The fact that the coefficient of kinship is the same for parent-offspring and full sibs points out the inadequacy of a single number to capture the genetic relatedness of relatives. A parent and its offspring always have exactly one allele each that are identical by descent.

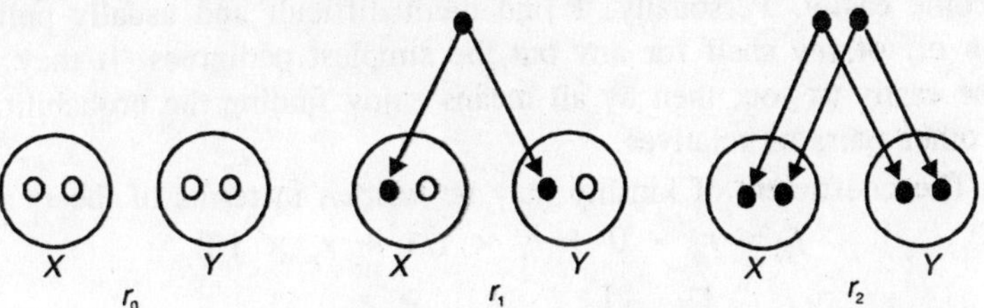

Fig. 6.2. The three possible numbers of shared alleles between relatives.

(The other allele in the offspring comes from the other parent.) Full sibs, on the other hand, may have zero, one, or two pairs of alleles that are identical by descent. Obviously, another measure of genetic relatedness is needed. The most complete measure is the set of probabilities of sharing zero, one, or two pairs of identical-by-descent alleles.

Table 6.1. The probability of two relatives sharing zero, one, or two alleles that are identical by descent.

Relationship	r_0	r_1	r_2
Parent-offspring	0	1	0
Full sibs	1/4	1/2	1/4
Half sibs	1/2	1/2	0
First cousins	3/4	1/4	0

The parent-offspring calculation of r_1 should be obvious. The full sib case will take a little more effort. The probability $r_2 = 1/4$ may be reasoned as follows. Pick one of the two alleles in individual *X*. The probability that an allele identical by descent with this allele is in individual *Y* is 1/2. Now pick the other allele in *X*. As it necessarily came from the other parent, its chance of having an identical-by-descent allele in *Y* is independent of the history of the first allele and is also 1/2. By independence, $r_2 = 1/4$. The converse of this argument gives $r_0 = 1/4$; $r_1 = 1/2$ is obtained by subtraction.

The difference in the two pairs of relatives, parent-offspring and full sibs, is now apparent. A parent and offspring always have one allele each that are identical by descent, while this occurs only one-half the time in full sibs. On the other hand, full sibs may share two or no identical-by-descent alleles, something that never happens with a parent and its offspring. As a consequence, in certain situations full sibs may appear more similar than parent-offspring, but only when there is dominance in the phenotype.

One needs a certain talent for the calculation of these probabilities to come easily. Personally, I find them difficult and usually pull a book off of my shelf for any but the simplest pedigrees. If they do come easily to you, then by all means enjoy finding the probabilities for other pairs of relatives.

The coefficient of kinship may be written in terms of the r_i as

$$f_{xy} = r_0 \times 0 + r_1 \times 1/4 + r_2 \times 1/2$$
$$= \frac{1}{4}r_1 + \frac{1}{2}r_2.$$

Each term in the sum corresponds to one of three mutually exclusive ways that two chosen alleles might be identical by descent. For example, the second term is the probability that *X* and *Y* share exactly one identical by descent allele times the probability that these two alleles are chosen, given that they are shared. The mean number of shared alleles is

$$\begin{aligned}\bar{r} &= 0 \times r_0 + 1 \times r_1 + 2 \times r_2 \\ &= r_1 + 2r_2\end{aligned}$$

One-half the mean number of shared alleles,

$$r = \bar{r}/2 = \frac{1}{2}r_1 + r_2, \qquad \text{...(1)}$$

is called the coefficient of relatedness, *r*. Some authors use the coefficient of relatedness where we use the coefficient of kinship. The two are related by

$$f_{xy} = \frac{1}{2}r.$$

INBREEDING

Inbreeding occurs when an individual mates with a relative. The progeny of such matings are more likely to be homozygous than are progeny produced by random matings. The level of inbreeding is measured by the inbreeding coefficient, F_I, which is the probability that the two alleles in an individual are identical by descent. As one of these alleles comes from one parent and the other from the other parent, the inbreeding coefficient of an individual is just the coefficient of kinship of its parents, $F_I = f_{xy}$. For example, the inbreeding coefficient of the offspring of a mating between full sibs is 1/4 because $f_{FS} = 1/4$.

As inbreeding increases the probability of being homozygous and as the average heterozygous effect of alleles affecting viability is less than one-half (at least in *Drosophila*), we would expect that inbred individuals would be less viable than outbred individuals. One demonstration is given, which graphs the viability of young human children as a function of their inbreeding coefficient. The functional dependence of viability on F_I may be described theoretically in much the same way that we described the effects of complete inbreeding in the Greenberg and Crow experiment. The particular approach that we will use comes from the classic paper by Newton Morton, James Crow, and H. J. Muller published in 1956. To derive the mean viability of the offspring of relatives, we require the frequencies of the three genotypes, A_1A_1, A_1A_2, and A_2A_2, among the offspring. First, consider the frequency of the A_1A_1 homozygotes. An offspring could be A_1A_1 for two reasons. Its two alleles could be identical by descent, which occurs with probability F_I, and they both could be A_1, which occurs with probability p (given that they are identical by descent). Thus, the probability of being A_1A_1 for this reason is pF_I. The second way to be an A_1A_1 is to have the two alleles not identical by descent but both A_1 anyway. The probability of this event is $(1 - F_I)p^2$. (If the individual does not have two identical-by-descent alleles, then it is as if the individual were the product of random mating.) Similar reasoning leads to the following:

Genotype:	A_1A_1	A_1A_2	A_2A_2
Frequency:	$p^2(1 - F_I) + pF_I$	$2pg(1 - F_I)$	$q^2(1 - F_I) + qF_I$
Fitness:	1	$1 - hs$	$1 - s$

The mean fitness of the offspring is

$$\bar{w} = 1 - 2pq(1 - F_I)hs - [q^2(1 - F_I) + qF_I]s$$
$$= 1 - a - bF_I,$$

where

$$a = 2pqsh + q^2s$$
$$b = 2pqs(1/2 - \text{h}).$$

In real populations, $h < 1/2$; therefore, b is always positive. Consequently, $\bar{w}$ is a decreasing linear function of F_I.

The probability of survival of *a* child, *S*, is affected by all of its loci plus various nongenetic factors,

$$S = \prod_j (1 - x_j) \prod_i (1 - a_i - b_i F_I),$$

where x_j represents the probability of death due to the *j*th nongenetic factor and the subscripts on *a* and *b* refer to the ith locus. The first product is over all nongenetic factors, and the second product is over all loci. By now, you should have recognized the implicit assumption that the loci act independently. This assumption was also invoked in the analysis of the Greenberg and Crow experiment. With our usual approximation for the product of numbers each of which is close to one, we get

$$S \approx e^{-A - BF_I}, \qquad \ldots(2)$$

where $\text{A} = \Sigma_j x_j + \Sigma_i a_i$ and $B = \Sigma_i b_i$.

As the natural logarithm of *S* is a linear function of F_I,

$$\log(S) = -A - BF_I,$$

it is a simple matter to estimate the values of A and B using standard regression methods. Morton, Crow, and Muller provide the following estimates:

$$A = 0.1612, B = 1.734.$$

The curve is plotted using these values of A and B. For small values of F_I,

$$S \approx e^{-A}(1 - BF_I),$$

which explains the roughly linear decrease of viability with F_I for the leftmost points.

The complete story about the relationship between mean fitness and the inbreeding coefficient is more complicated. The assumption that fitnesses at different loci may be multiplied is at variance with many experimental studies of inbreeding. Generally, the decrease in fitness with F_I goes faster than linear because of synergistic effects of mutations. Even the scant data bear this out as the rightmost two points lie below the curve drawn under the assumption of multiplicative epistasis. As this will be important in the discussion of the evolution of sex, we will return to the interaction between loci at that time. However, even when synergistic epistasis is present, the decrease in mean fitness with F_I is approximately linear for small F_I, as is the case with the human data.

Inasmuch as inbreeding increases homozygosity and, as a consequence, lowers fitness, it is not surprising that many species have evolved mechanisms to reduce the likelihood of mating with close relatives. Incest avoidance has been documented in many primate societies and, in humans, falls under the rubric of incest taboos, which are present in most cultures.

Now it is time to move from a description of the offspring of matings between relatives to a description of an inbred population. It should come as a pleasant surprise that there is nothing to do other than to note that F in the generalized Hardy-Weinberg is precisely the inbreeding coefficient of the population. In doing so, we imagine a population in which mating with relatives is common. We need not specify the kinds of matings that are taking place, as long as we happen to know F_I. Unfortunately, if we do know the sorts of matings, it is very difficult to find F_I for the population. For example, if we knew that one-half of the time individuals mated at random, one-quarter of the time they mated with their first cousins, one-eighth of the time with their grandparents, and the remaining times equally between full sibs, half sibs, and second cousins, it would be a struggle to determine F_I for the population.

The one example of inbreeding that is easy to study theoretically happens to be the one that is most relevant to natural populations, mixed selfing and outcrossing as practiced in many plant species. Under the mixed model, an individual is created by an act of selfing (literally, mating with one's self) with probability a or by random mating with probability $(1 - \alpha)$. Selfing will, in general, change F_I. The value of F_I in the next generation, as a function of its current value, is

$$F_I' = \alpha[F_I + (1 - F_I)\frac{1}{2}].$$

An individual in the next generation can have two alleles that are identical by descent only if the individual *is* produced by an act of selfing. Thus, the inbreeding coefficient in the next generation is multiplied by the probability of selfing, α. Given that an individual was produced by selfing, it can have two identical-by-descent alleles for one of two reasons: either its parent had two identical-by-descent alleles, which occurs with probability F_I, or its parent's alleles were not identical by descent, but the two alleles in the offspring are copies of the same allele in the parent, which occurs with probability $(1 - F_I)(1/2)$.

The change in F_I in a single generation is

$$\Delta_\alpha F_I = \frac{2-\alpha}{2}\left(\frac{\alpha}{2-\alpha} - F_I\right).$$

Partial selfing produces two antagonistic forces: selfing increases F_I while out-crossing decreases F_I. Eventually, an equilibrium is reached where $\Delta_\alpha F_I = 0$,

$$\hat{F}_I = \frac{\alpha}{2-\alpha}.$$

Notice that we are able to calculate F_I without making use of allele frequencies; we conclude immediately that allele frequencies do not change under mixed selfing and outcrossing. In fact, allele frequencies do not change under any system of inbreeding. Genotype frequencies do change, and they change in such a way that there are fewer heterozygotes than are seen in an outbreeding population.

Some interesting evolutionary questions arise with species that are capable of both selfing and outcrossing. For example, in many plant species there is an intrinsic advantage to selfing, which leads to the evolutionary conundrum: Why don't all plant species self? The

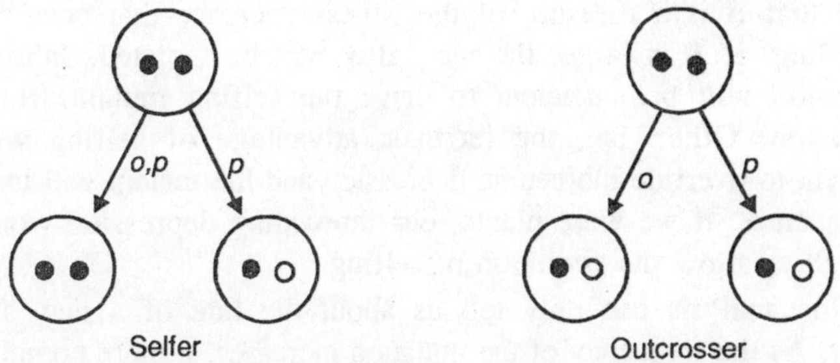

Fig. 6.3. The gametes produced by a selfer and an outcrosser.

outcrossing pedigree on the right represents a typical individual in an outcrossing population of constant size. This individual leaves behind, on average, two gametes, one carried in an ovule and the other in a pollen grain.

Suppose a mutant appears that self-fertilizes all of its ovules, as illustrated on the left side of the figure. Suppose also that there is enough pollen in each individual of this species that the few grains needed for self-pollination by the mutant represent a small fraction of the total pollen. As a consequence, the selfing mutant has essentially the same quantity of pollen available for outcrossing as does a nonselfing individual. All else being equal, the selfing mutant will leave behind three gametes for every two of the outcrossing plants. Two of the gametes are in its selfed offspring; one is in its outcrossed offspring. Thus, the mutant should increase in frequency, perhaps leading to the establishment of selfing as the usual mode of reproduction.

Of course, all else is not equal. The selfed offspring produced by the selfing mutant will be less fit due to inbreeding. The average number of gametes left behind by the selfer is

$$2e^{-A-BF_I}+e^{-A},$$

using Equation 2. (Each term in the sum is the number of gametes from the selfer times the fitness of the zygote containing the gametes.) The average number produced by an outbred individual is $2e^{-A}$. The selfing mutant will leave behind more gametes, on average, if

$$2e^{-A-BF_I} + e^{-A} > 2e^{-A},$$

or, because $F_I = 1/2$ for the selfed offspring of an outbred individual,

$$e^{-B/2} > 1/2.$$

Taking the natural logarithm of both sides and rearranging gives the condition for increase of the mutant,

$$B < 2\log(2) = 1.38.$$

Recall that B is a measure of the fitness decrease that occurs with inbreeding. If B is large, the inequality will be violated. Inbreeding depression will be sufficient to drive the selfing mutant from the population. Otherwise, the intrinsic advantage of selfing will be sufficient to override inbreeding depression and the mutant will increase in frequency. If we were plants, our inbreeding depression would be too high to allow the evolution of selfing.

This analysis can only tell us about the fate of a new selfing mutant. As the frequency of the mutation increases, a more complicated analysis is required to deal with all of the genotypes that are competing in the population. However, even this simple analysis suggests that plants should be either nearly complete selfers or nearly complete outcrossers. Intermediate levels of selfing will always be at a disadvantage when compared to higher levels of selfing when the inbreeding depression is small. Otherwise, more outcrossing is favoured. In fact, plants do tend to be at one extreme or the other. Inbreeding depression is clearly an important contributor to this pattern. Other factors are important as well. For example, weedy colonizing species might self as a way of guaranteeing that a mating occurs when the population density is very low. For a more thorough treatment of the evolution of self-fertilization, read the two papers by Russ Lande and Doug Schemske. The discussion of the evolution of selfing illustrates a style of inference based on the fate of rare mutants that is commonly used by evolutionists. A trait that is resistant to invasion by all mutants is often called an evolutionary stable strategy, or ESS. John Maynard Smith's book *Evolutionary Genetics* (1989) has many examples of the use of ESS methodology.

Thus far, we have stressed the deleterious effects of inbreeding. However, although the initiation of inbreeding is always detrimental, a population that maintains a steady level of inbreeding is not necessarily worse off than an out-breeding population. Inbreeding increases the effectiveness of selection against partially recessive ($h < 1/2$) deleterious alleles. At equilibrium, the frequency of these alleles will be lower in an inbreeding population than in an outbreeding one and, as a consequence, the genetic load will actually be lower than in the outbreeding population. To see this, we need to repeat the argument leading to $\Delta_s p$, but using the generalized Hardy-Weinberg frequencies. By now you should have little trouble showing that

$$\Delta_s p = (1 - F_I)\frac{pqs[ph + q(1-h)]}{\overline{w}} + F_I \frac{pqs}{\overline{w}},$$

where

$$\bar{w} = 1 - (1 - F_I)(2pqsh + q^2 s) - F_I qs.$$

When the frequency of the deleterious allele is very small,

$$\Delta_s p \approx (1 - F_I)qhs + F_I qs,$$

which is the inbreeding. The equilibrium between the increase of q by mutation ($\Delta_u p \approx -u$) and the decrease by selection occurs when

$$\hat{q}_I \approx \frac{u}{(1 - F_I)hs + F_I s}.$$

Comparing this, we see that inbreeding does lead to a decrease in the frequency of the deleterious allele at equilibrium because

$$(1 - F_I)hs + F_I s > hs$$

when $0 < h < 1$, and therefore $\hat{q}_I < \hat{q}$.

The genetic load of the equilibrium population is derived by first approximating the mean fitness by

$$\bar{w} \approx 1 - (1 - F_I)2qsh - F_I qs$$

and then substituting the equilibrium value $\hat{q}$ to obtain

$$\bar{w} \approx 1 - u\frac{2(1 - F_I)h + F_I}{(1 - F_I)h + F_I}.$$

Recall that the genetic load is, by definition,

$$L = \frac{w_{\max} - \bar{w}}{w_{\max}},$$

so that the equilibrium genetic load under selfing is

$$L \approx u\frac{2(1 - F_I)h + F_I}{(1 - F_I)h + F_I}.$$

With complete outbreeding, $F_I = 0$, $L = 2u$ just as we discovered earlier. However, when $F_I = 1$, the load is only u, one-half the outbred load. Thus, an equilibrium inbreeding population will have a lower genetic load than an outbreeding population! If, for some reason, an outcrossing population begins selfing, the mean fitness of the population will initially decrease because of inbreeding depression. Then, as selection begins to lower the frequency of deleterious mutations, the mean fitness will increase to a level that is higher than that of the original outbreeding population. This does not *mean* that the population is better off in a long-term evolutionary perspective. For example, highly inbred populations are less able to generate genetic diversity through recombination. An increase in genetic diversity can both speed up the rate of evolution and aid in the removal of deleterious mutations from chromosomes.

SUBDIVISION

Many species occupy such vast geographic areas or have such effective barriers to migration that they cannot behave as a single, randomly mating population. In such cases, there will be genetic differentiation between subpopulations, which leads to departures from Hardy-Weinberg for the entire species. In the case where there is random mating within each subdivision, the genotype frequencies for the entire species are described by a new incarnation of F called F_{ST}. For example, suppose a species is subdivided into two equal-sized patches with $p_1 = 1/4$ in the first patch and $p_2 = 3/4$ in the second. The genotype frequencies in the two patches are the Hardy-Weinberg frequencies given in the first two lines of the table below.

Genotype:	A_1	A_1A_1	A_1A_2	A_2A_2
Frequency in patch 1:	1/4	1/16	3/8	9/16
Frequency in patch 2:	3/4	9/16	3/8	1/16
Frequency in species:	1/2	5/16	3/8	5/16
Hardy-Weinberg frequencies:	1/2	1/4	1/2	1/4

The genotype frequencies for the entire species are the averages of the frequencies in the two patches. For example, the frequency of A_1A_1 is

$$\frac{1}{2}\times\frac{1}{16}+\frac{1}{2}\times\frac{9}{16}=\frac{5}{16},$$

as given in the third line of the table. The frequency of the A_1 allele in the entire species is obviously $p = 1/2$, so the expected Hardy-Weinberg frequencies are as given in the fourth line of the table. Even though each subpopulation has Hardy-Weinberg frequencies, the species as a whole does not. There are too many homozygotes; the excess requires that $F_{ST} = 1/4$, which is obtained by solving

$$2pq(1 - F) = (1 - F)/2 = 3/8.$$

As an observer, you have no way of knowing whether the excess of homozygotes is due to inbreeding, to subdivision, or to some other cause. You could sample the species more carefully until you identify the geographic substructure, but this is often more difficult than it first appears.

This example may be generalized to an arbitrary number of patches with any allele frequency in each. The aim is to find an expression for F_{ST} in terms of the allele frequencies in the subpopulations and the relative sizes of the subpopulations. Later, we will see how to use F_{ST} to investigate the role of migration in the genetic structure of a species.

Let p_i be the frequency of the A_1 allele in the ith subpopulation. Let the relative contribution of this subpopulation to the species or sample be c_i, $\Sigma c_i = 1$. Let p be the average frequency of the A_1 allele across patches, $p = \Sigma c p_i$, and let $q = 1 - p$. As with the example, the frequencies of genotypes are

Genotype:	A_1A_1	A_1A_2	A_2A_2
In *i*th patch	p_i^2	$2p_iq_i$	q_i^2
In species:	$\Sigma c p_i^2$	$\Sigma c_i 2p_iq_i$	$\Sigma c_i q_i^2$
In species:	$p^2(1 - F_{ST}) + pF_{ST}$	$2pq(1 - F_{ST})$	$q^2(1 - F_{ST}) + qF_{ST}$

By equating the two ways of writing heterozygote frequencies in the species,

$$2pq(1-F_{ST})\sum c_i 2p_iq_i,$$

we get

$$F_{ST} = \frac{2pq - \Sigma c_i 2p_iq_i}{2pq}.$$

Using

$$2pq = 1 - p^2 - q^2$$

and

$$\sum c_i 2p_iq_i = 1 - \sum c_i p_i^2 - \sum c_i q_i^2 = -\sum c_i (p_i^2 + q_i^2),$$

the expression for F_{ST} may be written as

$$F_{ST} = \frac{\Sigma c_i (p_i^2 + q_i^2) - p^2 - q^2}{2pq} = \frac{G_S - G_T}{1 - G_T}, \qquad ...(3)$$

where

$$G_T = p^2 + q^2$$

$$G_S = \sum c_i (p_i^2 + q_i^2). \qquad ...(4)$$

G_T is the probability that two alleles drawn at random (with replacement) from the entire species are identical by state. G_S is the probability that two alleles drawn at random (with replacement) from a randomly chosen subdivision are identical by state. (With probability c_i the *i*th patch is chosen. Given this, with probability $p_i^2 + q_i^2$ two randomly drawn alleles are identical by state.) F_{ST} is seen as a measure of the difference between the probability that two alleles drawn from within a subdivision are identical compared to the probability that two alleles drawn at random from the species are identical.

F_{ST} may also be written as a function of the variance of the allele frequencies across patches by substituting

$$\sum c_i p_i^2 - p^2 = \mathrm{Var}\{p_i\}$$

and

$$\sum c_i q_i^2 - q^2 = \mathrm{Var}\{q_i\}$$

into the middle first expression for F_{ST} in Equation 3. This gives

$$F_{ST} = \frac{\mathrm{Var}\{p_i\} + \mathrm{Var}\{q_i\}}{1 - G_T}$$

$$= \frac{2\mathrm{Var}\{p_i\}}{1 - G_T}.$$

The final step follows from $q_i = 1 - p_i$. This result shows that F_{ST} is always positive. As long as there is variation in the allele frequency across subdivisions ($\mathrm{Var}\{p_i\} > 0$), the genotype frequencies of the species will exhibit a deficiency of heterozygotes, a condition known as Wahlund's effect.

Inbreeding and subdivision both lead to an apparent deficiency of heterozygotes over Hardy-Weinberg expectations. Thus, nothing can be said about the causes of an observed deficiency without more information. One interesting case where the deficiency of heterozygotes could be attributed to subdivision involves limpets from the intertidal zone of Western Australia as reported in two 1984 papers by Michael Johnson and Robert Black. An electrophoretic survey of protein polymorphism in limpets from the intertidal zone found a deficiency of heterozygotes with F_{ST} values ranging as high as 0.018. A careful study revealed that there was temporal variation in allele frequencies of new recruits. Thus, the limpets at any one spot in the intertidal zone represented a mixture of groups of limpets with different allele frequencies, with a consequent Wahlund's effect.

F_{ST} is a very crude measure of the geographic structure of a species. As it is just a single number, it cannot distinguish between, say, an allele frequency dine from north to south and a hodgepodge of independent allele frequencies in each subdivision. Both patterns produce a positive F_{ST}. There is one very special situation where F_{ST} may tell something about the migration rates of species. This occurs when the segregating alleles are neutral and there are a large number of subdivisions with equal migration rates between them. Sewall Wright was the first to explore this model, which is usually called the island model. As we describe the island model, it will not appear to apply to a large number of subdivisions. Bear with me; we will get to that model after a short development of the literal island model.

There is a very large, effectively infinite, mainland population that sends migrants to the island. There are N diploid individuals living on the island. Each generation a fraction, m, of these individuals depart for places unknown, to be replaced by individuals from the mainland population. For simplicity, we will assume that haploid gametes rather than diploid individuals migrate. While not realistic for many species, this assumption has no effect on the properties of the island model that are of interest. Thus, $2Nm$ gametes migrate each generation.

Genetic drift and migration are the only evolutionary forces acting on the island population. Genetic drift works to eliminate genetic variation on the island, while migration brings in new genetic variation from the mainland. Eventually, an equilibrium is reached with the level of variation on the island determined by N, m, and the amount of variation in the mainland population. There should be a familiar ring to this model. If we were to substitute *mutation* for *migration*, the island model would sound remarkably like the model of mutation and drift,

$$\hat{G} = \frac{1}{1+4Nu}.$$

In fact, with an appropriate assumption, the equilibrium level of variation in the island is

$$\hat{G} = \frac{1}{1+5Nm}, \qquad ...(5)$$

The appropriate assumption is simply that each migrant allele is genetically unique. If the mainland population is truly infinite, then this assumption is automatically satisfied as $\hat{G} = 0$ when $N = \infty$. Thus, the change in H on the island due to migration is

$$\Delta_m H = 2m(1 - H),$$

which is the migration equivalent for mutation. The change in H due to genetic drift is

$$\Delta_N H = -\frac{1}{2N} H.$$

At equilibrium, when $\Delta_m H + \Delta_N H = 0$, $\hat{G}$ is given by Equation 5.

The homozygosity, as measured by G, changes fairly quickly from near one to near zero as $4Nm$ changes from being less than one to greater than one. When near zero, genetic drift removes most of the variation from the island. When large, the island population's homozygosity is like that of the mainland. Thus, when the fraction of migrants is greater than about $1/(4N)$, the effects of isolation become unimportant and the island appears to be infinite in size, as indicated by its value of G. This observation becomes compelling when expressed in numbers of migrants rather than in fraction of migrants. Isolation disappears if $m > 1/4N$, or, equivalently, if $2Nm > 1/2$. The absolute number of diploid migrants each generation is $2Nm$ (strictly, the number of pairs of haploid genomes, as gametes rather than zygotes migrate in our model). If more than one individual migrates every other generation, then the effects of isolation become unimportant. Surprisingly, this statement is independent of the population size of the island. One might have thought that more migration would be required to make large islands like the mainland. However, in large islands drift is a weak force, so less migration is needed to balance drift. The message is clear: very little migration can make a subdivided species appear like one large randomly mating species when neutral alleles are involved.

It is now time to bring F_{ST} back into the discussion and with it a change in the description of the island model. Instead of a mainland-island structure, imagine a large number of patches, each exchanging a fraction, m, of their gametes with all of the other patches each generation.. Said another way, the $2Nm$ immigrants into each patch each generation are chosen at random from the other patches. If the number of patches increases to infinity, the species population size will be infinite, while the size of each patch will remain fixed at N. From the point of view of a particular patch, the rest of the patches collectively are just like the mainland in Wright's island model.

The homozygosity in each patch is still given by Equation 5, which is essentially the same as G_S as defined in Equation 4. (The qualification "essentially" refers to an error of approximately $1/2N$ that occurs when the expected value of the homozygosity, G, is equated with the probability of identity by state, G.) G_S is the average value of the homozygosity across subdivisions. Equation 3 for F_{ST} also requires G_T, which is the probability that two alleles drawn at random from the species are identical by state. As the species population size is infinite, $G_T = 0$. Thus, for the island model,

$$F_{ST} = \frac{G_S - G_T}{1 - G_T} = G_S = \frac{1}{1 + 4Nm}.$$

If we knew the patch size for a species, *N,* and its F_{ST}, then we would know its migration rate, assuming that all of the assumptions of the model are met.

This is a big *if.* To use Wright's island model to estimate *m* or *mN,* one must be certain that the variation at the locus under study is neutral, that the population is at equilibrium, and that the migration pattern at least approximates that of the island model. Despite the fact that these assumptions are seldom, if ever, met, F_{ST} is frequently used to gain some insights into the genetic structure of a species. For a discussion of the pitfalls of this approach, read the paper by Monty Slatkin and Nick Barton (1989).

7

Process that Change Allelic Frequencies

We continue our discussion of the genetics of the evolutionary process. This chapter is devoted to a discussion of some of the effects of violating, or relaxing, the assumptions of the Hardy-Weinberg equilibrium other than random mating. Here we consider the effects of mutation, migration, small population size, and natural selection on the Hardy-Weinberg equilibrium. These processes usually change allelic frequencies.

Models for Population Genetics

The steps we need to take to solve for equilibrium in population genetics models follow the same general pattern regardless of what model we are analyzing. We emphasize that these models were developed to help us understand the genetic changes taking place in a population. The models shed light on nonintuitive processes and help quantify intuitive processes. The steps in the models can be outlined as follows:

1. Set up an algebraic model.
2. Calculate allelic frequency in the next generation, q_{n+1}.
3. Calculate change in allelic frequency between generations, Δq.
4. Calculate the equilibrium condition, $\hat{q}$ (q-hat), at $\Delta q = 0$.
5. Determine, when feasible, if the equilibrium is stable.

Mutation

Mutational Equilibrium

Mutation affects the Hardy-Weinberg equilibrium by changing one allele to another and thus changing allelic and genotypic frequencies.

Consider a simple model in which two alleles, *A* and *a,* exist. *A* mutates to *a* at a rate of μ, (mu), and *a* mutates back to *A* at a rate of ν *(nu):*

$$A \underset{\nu}{\overset{\mu}{\rightleftarrows}}$$

If p_n is the frequency of *A in* generation *n* and q_n is the frequency of *a* in generation *n,* then the new frequency of *a,* $q_{n\pm1}$, is the old frequency of *a* plus the addition of *a* alleles from forward mutation and the loss of *a* alleles by back mutation. That is,

$$q_{n+1} = q_n + \mu p_n - \nu q_n \qquad ...(1)$$

in which μp_n is the increment of *a* alleles added by forward mutation, and νq_n is the loss of *a* alleles due to back mutation. Equation 1 takes into account not only the rate of forward mutation, μ, but also p_n, the frequency of *A* alleles available to mutate. Similarly, the loss of *a* to *A* alleles is the product of both the rate of back mutation, ν, and the frequency of the *a* allele, q_n. Equation 1 completes the second modeling step, derivation of an expression for $q_{n\pm1}$, allelic frequency after one generation of mutation pressure. The third step is to derive an expression for the change in allelic frequency between two generations. This change (Δq) is simply the difference between the allelic frequency at generation $n + 1$ and the allelic frequency at generation *n*. Thus, for the *a* allele

$$\Delta q = q_{n+1} - q_n = (q_n + \mu p_n - \nu q_n) - q_n \qquad ...(2)$$

which simplifies to

$$\Delta q = \mu p_n - \nu q_n \qquad ...(3)$$

The next step in the model is to calculate the equilibrium condition $\hat{q}$ or the allelic frequency when there is no change in allelic frequency from one generation to the next—that is, when Δq (equation 3) is equal to zero:

$$\Delta q = \mu p_n - \nu q_n = 0 \qquad ...(4)$$

Thus,

$$\mu p_n = \nu q_n \qquad ...(5)$$

Then, substituting $(1 - q_n)$ for p_n (since $p = 1 - q$), gives

$$\mu(1 - q_n) = \nu q_n$$

or, by rearranging:

$$\hat{q} = \frac{\mu}{\mu + \nu} \qquad ...(6)$$

And, since $p + q = 1$,

$$\hat{p} = \frac{\nu}{\mu + \nu} \quad ...(7)$$

We can see from equations 6 and 7 that an equilibrium of allelic frequencies does exist. Also, the equilibrium value of allele a ($\hat{q}$) is directly proportional to the relative size of μ, the rate of forward mutation toward a. If $\mu = \nu$, the equilibrium frequency of the a allele ($\hat{q}$) will be 0.5. As μ, gets larger, the equilibrium value shifts toward higher frequencies of the a allele.

Stability of Mutational Equilibrium

Having demonstrated that allelic frequencies can reach an equilibrium due to mutation, we can ask whether the mutational equilibrium is stable. A stable equilibrium is one that returns to the original equilibrium point after being perturbed. An unstable equilibrium is one that will not return after being perturbed but, rather, continues to move away from the equilibrium point. As we mentioned in the last chapter, the Hardy-Weinberg equilibrium is a neutral equilibrium: It remains at the allelic frequency it moved to when perturbed.

Stable, unstable, and neutral equilibrium points can be visualized as marbles in the bottom of a concave surface (stable), on the top of a convex surface (unstable), or on a level plane (neutral). Although more sophisticated mathematical formulas exist for determining whether an equilibrium is stable, unstable, or neutral, we will use graphical analysis for this purpose.

The process of graphical analysis, which provides an understanding of the dynamics of an event or process by representing the event in graphical form. We have graphed equation 3, the Δq equation of mutational dynamics. The ordinate, or y-axis, is Δq, the change in allelic frequency. The abscissa, or x-axis, is q, or allelic frequency. The diagonal line is the Δq equation, the relationship between Δq and q. Note that Δq can be positive *(q is increasing)* or negative *(q is decreasing)*, whereas q is always positive (0-1.0). Graphical analysis can provide insights into the dynamics of many processes in population genetics.

The diagonal line crosses the $\Delta q = 0$ line at the equilibrium value ($\hat{q}$) of 0.167. This line also shows us the changes in allelic frequency that occur in a population not at the equilibrium point. We will look at two examples of populations under the influence of mutation pressure, but not at equilibrium: one at $q = 0.1$ (below equilibrium) and one at $q = 0.9$ (above equilibrium).

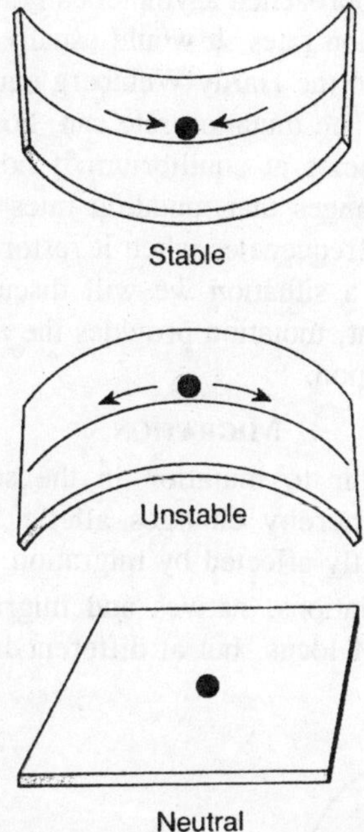

Fig. 7.1. Types of equilibria: stable, unstable, and neutral.

If we substitute $q = 0.1$ into equation 3, we get a Δq value of 4×10^{-6}. If we substitute $q = 0.9$ into the equation, we get a Δq value of -4.4×10^{-5}. In other words, when the population is below equilibrium, q increases ($\Delta q = +4 \times 10^{-6}$); if the population is above equilibrium, q decreases ($\Delta q = -4.4 \times 10^{-5}$). We can read these same conclusions directly from the graph.

We can see that the mutational equilibrium is a stable one. Any population whose allelic frequency is not at the equilibrium value tends to return to that equilibrium value. A shortcoming of this model is that it provides no obvious information revealing the time frame for reaching equilibrium. To derive the equations needed to determine this parameter is beyond our scope. (We could use computer simulation or integrate equation 3 with respect to time.) In a large population, any great change in allelic frequency caused by mutation pressure alone takes an extremely long time. Most mutation rates are on the order of 10^{-5}, and equation 3 shows that change will be very slow with values of this magnitude. For example, if $\mu = 10^{-5}$, $v = 10^{-6}$, and $p = q = 0.5$, $\Delta q = (0.5 \times 10^{-5}) - (0.5 \times 10^{-6}) = 4.5 \times 10^{-6}$, or 0.0000045. It usually takes thousands of generations to get near equilibrium, which is approached asymptotically. As you can see from the low values of mutation rates, it would usually be nearly impossible to detect perturbations to the Hardy-Weinberg equilibrium by mutation in any one generation. The mutation rate can, however, determine the eventual allelic frequencies at equilibrium if no other factors act to perturb the gradual changes that mutation rates cause. Mutation can also affect final allelic frequencies when it restores alleles that natural selection is removing, a situation we will discuss at the end of the chapter. More important, mutation provides the alternative alleles that natural selection acts upon.

MIGRATION

Migration is similar to mutation in the sense that it adds or removes alleles and thereby changes allelic frequencies. Human populations are frequently affected by migration.

Assume two populations, natives and migrants, both containing alleles *A* and *a* at the *A* locus, but at different frequencies (p_N and q_N

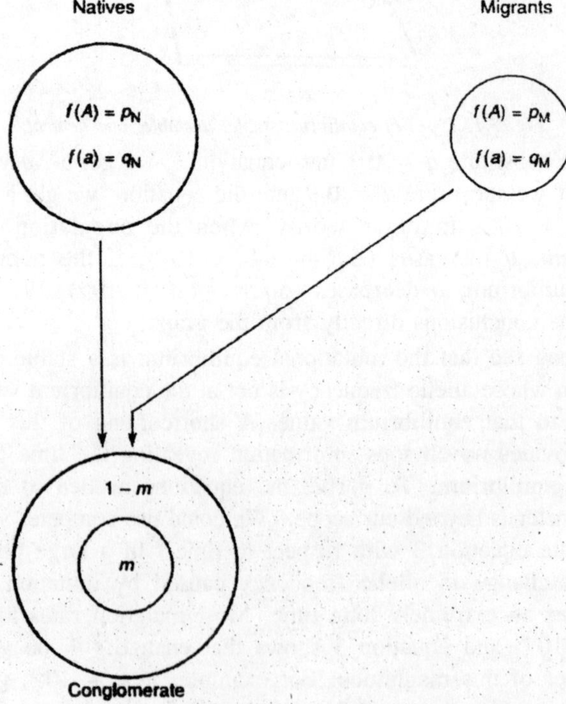

Fig. 7.2. Diagrammatic view of migration.

versus p_M and q_M). Assume that a group of migrants joins the native population and that this group of migrants makes up a fraction *m* (e.g., 0.2) of the new conglomerate population. Thus, the old residents, or natives, will make up a proportionate fraction (1 – *m;* e.g., 0.8) of the combined population. The conglomerate *a*-allele frequency, q_c, will be the weighted average of the allelic frequencies of the natives and migrants:

$$q_c = mq_M + (1 - m)q_N \quad ...(8)$$

$$q_c = q_N + m(q_M - q_N) \quad ...(9)$$

The change in allelic frequency, *a,* from before to after the migration event is

$$\Delta q = q_c - q_N = [q_N + m(q_M - q_N)] - q_N \quad ...(10)$$

$$\Delta q = m(q_M - q_N) \quad ...(11)$$

We then find the equilibrium value, $\hat{q}$ (at $\Delta q = 0$). Remembering that, in a product series, any multiplier with the value of zero makes the whole expression zero, Δq will be zero when either

$$m = 0 \text{ or } q_M - q_N = 0; \; q_M = q_N$$

The conclusions we can draw from this model are intuitive. Migration can upset the Hardy-Weinberg equilibrium. Allelic frequencies in a population under the influence of migration will not change if either the size of the migrant group drops to zero *(m,* the proportion of the conglomerate made up of migrants, drops to zero) or the allelic frequencies in the migrant and resident groups become identical.

This migration model can be used to determine the degree to which alleles from one population have entered another population. It can analyze the allele interactions in any two populations. We can, for example, analyze the amount of admixture of alleles from Mongol populations with eastern European populations to explain the relatively high levels of blood type B in eastern European populations (if we make the relatively unrealistic assumption that each of these groups is homogeneous). The calculations are also based on a change happening all in one generation, which did not happen. Blood type and other loci can be used to determine allelic frequencies in western European, eastern European, and Mongol populations. We can rearrange equation 20.9 to solve for *m,* the proportion of migrants:

$$m = \frac{q_c - q_N}{q_M - q_N} \quad ...(12)$$

From one sample, we find that the *B* allele is 0.10 in western Europe, taken as the resident or native population (q_N); 0.12 in eastern Europe, the conglomerate population (q_c); and 0.21 in Mongols, the migrants (q_M) . Substituting these values into equation 12 gives a value for *m* of 0.18. That is, given the stated assumptions, 18% of the alleles in the eastern European population were brought in by genetic mixture with Mongols.

When a migrant group first joins a native group, before genetic mixing (mating) takes place, the Hardy-Weinberg equilibrium of the conglomerate population is perturbed, even though both subgroups are themselves in Hardy-Weinberg proportions. A decrease will occur in heterozygotes in the conglomerate population as compared to what we would predict from the allelic frequencies of that population (the average allelic frequencies of the two groups). This is a phenomenon of subdivision referred to as the *Wahlund effect*. The reason this happens is because the relative proportions of heterozygotes increase at intermediate allelic frequencies. As allelic frequencies rise above or fall below 0.5, the relative proportion of heterozygotes decreases.

In a conglomerate population, the allelic frequencies will be intermediate between the values of the two subgroups because of averaging. This generally means the predicted proportion of heterozygotes will be higher than the actual average proportion of heterozygotes in the two subgroups. Assume that the two subgroups each make up 50% of the conglomerate population. In subgroup 1, $p = 0.1$ and $q = 0.9$; in subgroup 2, $p = 0.9$ and $q = 0.1$. Each subgroup will have 18% heterozygotes. The average, $(0.18 + 0.18)/2 = 0.18$, is the proportion of heterozygotes actually in the population. However, the conglomerate allelic frequencies are $p = 0.5$ and $q = 0.5$, leading to the expectation that 50% of the population will be heterozygotes. Hence, the observed frequency of heterozygotes is lower than the expected frequency (i.e., the Wahlund effect).

Table 7.1. The Wahlund effect: Heterozygote frequencies are below expected in a conglomerate population

	Subgroup 1	*Subgroup 2*	*Conglomerate*	
p	0.1	0.9	0.5	
q	0.9	0.1	0.5	
			Expected	Observed
p^2	0.01	0.81	0.25	0.41
$2pq$	0.18	0.18	0.50	0.18
q^2	0.81	0.01	0.25	0.41

We should note that the same logic holds even if both populations have allelic frequencies above or below 0.5. Also, this effect happens when an observer samples what he or she thinks is a single population but is actually a population subdivided into several denies. When most population geneticists sample a population and find a deficiency of heterozygotes, they first think of inbreeding and then of subdivision, the Wahlund effect. (A further complication is that inbreeding leads to subdivision, and subdivision leads to inbreeding. Statistics have been developed to try to separate the effects of these two phenomena.) As soon as random mating occurs in a subdivided population, Hardy-Weinberg equilibrium is established in one generation. We refer to a population in which the individuals are mating at random as unstructured or *panmictic*.

Small Population Size

Another variable that can upset the Hardy-Weinberg equilibrium is small population size. The Hardy-Weinberg equilibrium assumes an infinitely large population because, as defined, it is *deterministic*, not stochastic. That is, the Hardy-Weinberg equilibrium predicts exactly what the allelic and genotypic frequencies should be after one generation; it ignores variation due to sampling error. Obviously, every population of organisms on earth violates the Hardy-Weinberg assumption of infinite population size.

Sampling Error

The zygotes of every generation are a sample of gametes from the parent generation. Sampling errors are the changes in allelic frequencies from one generation to the next that are due to inexact sampling of the alleles of the parent generation. Toss a coin one hundred times, and chances are, it will not land heads exactly fifty times. However, as the number of coin tosses increases, the percentage of heads will approach 50%, a percentage reached with certainty only after an infinite number of tosses. The same applies to any sampling problem, from drawing cards from a deck to drawing gametes from a gene pool.

If small population size is the only factor causing deviation from Hardy-Weinberg equilibrium, it will cause the allelic frequencies of a population to fluctuate from generation to generation in the process known as random genetic drift. In other words, an *Aa* heterozygote will sometimes produce several offspring that have only the *A* allele, or sometimes random mortality will kill a disproportionate number of *as* homozygotes. In either case, the next generation may not have the same allelic frequencies as the present generation. The end result will be either fixation or loss of any given allele ($q = 1$ or $q = 0$), although which will be fixed or lost depends on the original allelic frequencies. The rate of approach to reach the fixation-loss endpoint depends on the size of the population.

Simulation of Random Genetic Drift

We can investigate the process of random genetic drift mathematically by starting with a large number of populations of the same finite size and observing how the distribution of allelic frequencies among the populations changes in time due only to random genetic drift. For example, we can start with one thousand hypothetical populations, each containing one hundred individuals, with the frequency of the *a* allele, q, 0.5 in each. We measure time in generations, t, as a function of the population size, N (one hundred in this example). For instance, $t = N$ is generation one hundred, $t = N/5$ is generation twenty, and $t = 3N$ is generation three hundred. Then, by using computer simulation (or the *Fokker-Planck equation*, which physicists use to describe diffusion processes such as Brownian motion), we generate the series of curves. These curves show that as the number of generations increases, the populations begin to diverge from $q = 0.5$. Approximately the same number of populations go to q values above 0.5 as go to q values below 0.5. Therefore, the distribution spreads symmetrically. When the distribution of allelic frequencies reaches the sides of the graph, some populations become fixed for the *a* allele and some lose it. In a sense, the sides act as sinks: Any population that has the *a* allele lost or fixed will be permanently removed from the process of random genetic drift. Without mutation to bring one or the other allele back into the gene pool, these populations maintain a constant allelic frequency of zero or 1.0

At a point between N (one hundred) and $2N$ (two hundred) generations, the distribution of allelic frequencies flattens out and begins to lose populations to the edges (fixation or loss) at a constant rate. The rate of loss is about $1/2N$ (1/200), or 0.5% of the populations per generation. If the initial allelic frequency was not 0.5, everything is shifted in the distribution, but the basic process is the same—in all populations, sampling error causes allelic frequencies to drift toward fixation or elimination. If no other factor counteracts this drift, every population is destined to eventually be either fixed for or deficient in any given allele.

The amount of time the process takes depends on population size. The example used here was based on small populations of one hundred. If we substitute one million for one hundred, a flat distribution of populations would not be reached for two million generations, rather than two hundred generations. Thus, a population experiences the effect of random genetic drift in inverse proportion to its size: Small populations rapidly fix or lose a given allele, whereas large populations take longer to show the same effects. Genetic drift also shows itself in several other ways.

Founder Effects and Bottlenecks

Several well-known genetic phenomena are caused by populations starting at or proceeding through small numbers. When a population is initiated by a small, and therefore genetically unrepresentative, sample of the parent population, the genetic drift observed in the subpopulation is referred to as a *founder effect*. A classic human example is the population founded on Pitcairn Island by several of the *Bounty* mutineers and some Polynesians. The unique combination of Caucasian and Polynesian traits that characterizes today's Pitcairn Island population resulted from the small number of founders for the population.

Sometimes populations go through *bottlenecks*, periods of very small population size, with predictable genetic results. After the bottleneck, the parents of the next generation have been reduced to a small number and may not be genetically representative of the original population. The field mice on Muskeget Island, Massachusetts, have a white forehead blaze of hair not commonly found in nearby mainland populations. Presumably, the island population went through a bottleneck at the turn of the century, when cats on the island reduced the number of mice to near zero. The population was reestablished by a small group of mice that happened by chance to contain several animals with this forehead blaze.

Natural Selection

Although mutation, migration, and random genetic drift all influence allelic frequencies, they do not necessarily produce populations of individuals that are better adapted to their environments. Natural selection, however, tends to that end. The consequence of natural selection, Darwinian evolution, is considered in detail in the next chapter. We discuss here the algebra behind the process of natural selection. Artificial selection, as practiced by animal and plant breeders, follows the same rules.

How Natural Selection Acts

Selection, or *natural selection*, is a process whereby one phenotype and, therefore, one genotype leaves relatively more offspring than another genotype, measured by both reproduction and survival. Selection is thus a matter of *reproductive success*, the relative contribution of that genotype to the next generation. It is important to remember that selection acts on whole organisms and thus on phenotypes. However, we analyze the process by looking directly at the genotype, usually only at one locus.

Fitness

A measure of reproductive success is the *fitness*, or *adaptive value*, of a genotype. A genotype that, compared with other genotypes, leaves relatively more offspring that survive to reproduce has the higher fitness. (Note that this use of the word *fitness* differs from our common notion of physical fitness.)

Fitness is usually computed to vary from zero to one (0-1) and is always related to a given population at a given time. For example, in a normal environment, fruit flies with long wings may be more fit than fruit flies with short wings. But in a very windy environment, a fruit fly with limited flying ability may survive better than one with the long-winged genotype, which will be blown around by the wind. Thus, fitness (usually assigned the letter *W*) is relative to a given circumstance. In a given environment, the genotype that leaves the most offspring is usually assigned a fitness of $W = 1$, and a lethal genotype has a fitness of $W = 0$. Any other genotype has a fitness value between zero and one. A number of factors can decrease this fitness value, *W*, below one. A *selection coefficient* measures the sum of forces acting to prevent reproductive success. It is usually represented by the letter *s* or *t* and is defined by the fitness equation

$$W = 1 - s \quad ...(13)$$

and

$$s = 1 - W \quad ...(14)$$

Thus, as the selection coefficient increases, fitness decreases, and vice versa.

Components of fitness

Natural selection can act at any stage of the life cycle of an organism. It usually acts in one of four ways. (1) The reproductive success of a genotype can be affected by prenatal, juvenile, or adult survival. Differential survival of genotypes is referred to as viability selection or *zygotic selection*. (2) A heterozygote can produce gametes with differential success when one of its alleles fertilizes more often than the other. This is termed *gametic selection*. A well-studied case is the *t*-allele (tailless) locus in house mice; the gametes of as many as 95% of the heterozygous males of the *Tt* genotype carry the *t* allele. (This phenomenon is also referred to as *segregation distortion* or *meiotic drive*.) Selection can also take place in two areas of the reproductive segment of an organism's life cycle. (3) Some genotypes may mate more often than others (have greater mating success), resulting in *sexual selection*. Sexual selection usually occurs when members of the same sex compete for mates or when females have some form of choice. Adaptations for fighting, such as antlers in male elk, or displaying, such as the peacock's tail, are the results of sexual selection. (4) Finally, some genotypes may be more fertile than other genotypes, resulting in *fecundity selection*. The particular variable of the life cycle that selection acts upon is termed a *component of fitness*.

Effects of selection

Directional selection works by continuously removing individuals from one end of the phenotypic (and therefore, presumably, genotypic) distribution (e.g., short-necked giraffes are removed). Removal means

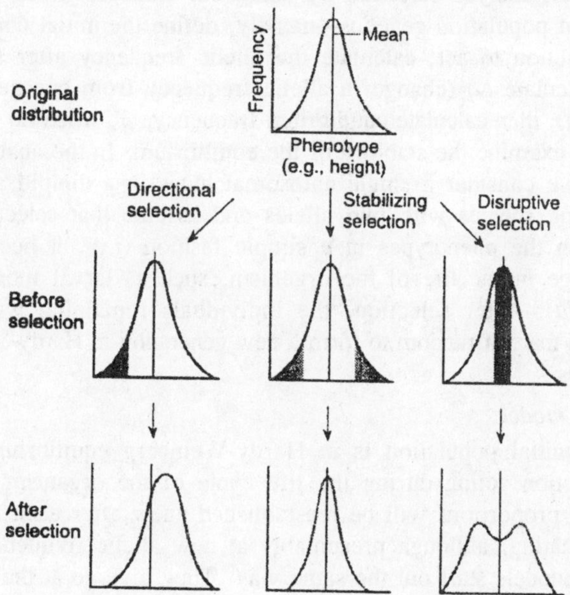

Fig. 7.3. Directional, stabilizing, and disruptive selection.

disappearance through death or failure to reproduce (genetic death). Thus, the mean is constantly shifted toward the other end of the phenotypic distribution; in our example, the mean shifts toward long-necked giraffes. The evolution of neck length in giraffes, presumably by directional selection, has been documented from the geologic record.

Stabilizing selection works by constantly removing individuals from both ends of a phenotypic distribution, thus maintaining the same mean over time. Stabilizing selection now works on giraffe neck length—it is neither increasing nor decreasing. ***Disruptive selection*** works by favouring individuals at both ends of a phenotypic distribution at the expense of individuals in the middle. It, like stabilizing selection, should maintain the same mean value for the phenotypic distribution. Disruptive selection has been carried out successfully in the laboratory for bristle number in *Drosophila*. Starting with a population with a mean number of sternopleural chaeta (bristles on one of the body plates) of about eighteen, investigators succeeded after twelve generations of getting a fly population with one peak of bristle numbers at about sixteen and another at about twenty-three.

Selection Against the Recessive Homozygote

We can analyze selection by using our standard model-building protocol of population genetics—namely, define the initial conditions; allow selection to act; calculate the allelic frequency after selection (q_{n+1}); calculate Δq (change in allelic frequency from one generation to the next); then calculate equilibrium frequency, $\hat{q}$, when Δq becomes zero; and examine the stability of the equilibrium. In the analysis that follows, we consider a single autosomal locus in a diploid, sexually reproducing species with two alleles and assume that selection acts directly on the phenotypes in a simple fashion (i.e., it occurs at a single stage in the life of the organism, such as larval mortality in *Drosophila*). After selection, the individuals remaining within the population mate at random to form a new generation in Hardy-Weinberg proportions.

Selection model

The initial population is in Hardy-Weinberg equilibrium. Even with selection acting during the life cycle of the organism, Hardy-Weinberg proportions will be reestablished anew after each round of random mating, although presumably at new allelic frequencies. All selection models start out the same way. They diverge at the point of assigning fitness, which depends on the way natural selection is acting. In the model, the dominant homozygote and the heterozygote have the same fitness ($W = 1$). Natural selection cannot differentiate between the two genotypes because they both have the same phenotype. The recessive homozygote, however, is being selected against, which means that it has a lower fitness than the two other genotypes ($W = 1 - s$).

After selection, the ratio of the different genotypes is determined by multiplying their frequencies (Hardy-Weinberg proportions) by their fitnesses. The procedure follows from the definition of fitness, which in this case is a relative survival value. Thus, only $1 - s$ of the *aa* genotype survives for every one of the other two genotypes. For example, if s were 0.4, then the fitness of the *aa* type would be $1 - s$, or 0.6. For every ten *AA* and *Aa* individuals that survive to reproduce, only six *aa* individuals would survive to reproduce. The total of the three genotypes after selection is $1 - sq^2$. This is,

$$p^2 + 2pq + q^2(1 - s) = p^2 + 2pq + q^2 - sq^2 = 1 - sq^2$$

Mean fitness of a population

The value $(1 - sq^2)$ is referred to as the *mean fitness of the population*, $\bar{W}$, because it is the sum of the fitnesses of the genotypes multiplied (weighted) by the frequencies at which they occur. Thus, it is a weighted mean of the fitnesses, weighted by their frequencies. The new ratios of the three genotypes can be returned to genotypic frequencies by simply dividing by the mean fitness of the population, $\bar{W}$. (Remember that a set of numbers can be converted to proportions of unity by dividing them by their sum.) The new genotypic frequencies are thus the products of their original frequencies times their fitnesses, divided by the mean fitness of the population.

After selection, the new allelic frequency (q_{n+1}) is the proportion of *aa* homozygotes plus half the proportion of heterozygotes, or

$$q_{n+1} = \frac{q^2(1-s)}{1-sq^2} + \frac{pq}{1-sq^2}$$

$$= \frac{q(q - sq + p)}{1-sq^2}$$

$$= \frac{q(1-sq)}{1-sq^2} \qquad \text{...(15)}$$

This model can be simplified somewhat if we assume that the *aa* genotype is lethal. Its fitness would be zero, and s, the selection coefficient, would be one. Equation 15 would then change to

$$q_{n+1} = \frac{q(1-q)}{1-q^2} \qquad \text{...(16)}$$

Since $(1 - q^2)$ is factorable into $(1 - q)(1 + q)$, equation 16 becomes

$$q_{n+1} = \frac{q(1-q)}{(1-q)(1+q)}$$

$$= \frac{q}{(1+q)} \quad \text{...(17)}$$

The change in allelic frequency is then calculated as

$$\Delta q = q_{n+1} - q = \frac{q}{1+q} - q$$

To solve this equation, q is multiplied by $(1 + q)/(1 + q)$ so that both parts of the expression are over a common denominator:

$$\Delta q = \frac{q - q(1+q)}{1+q}$$

$$= \frac{-q^2}{1+q} \quad \text{...(18)}$$

This is the expression for the change in allelic frequency caused by selection. Since selection will not act again until the same stage in the life cycle during the next generation, equation 18 is also an expression for the change in allelic frequency between generations.

Two facts should be apparent from equation 18. First, the frequency of the recessive allele *(q)* is declining, as indicated by the negative sign of the fraction. This fact should be intuitive because of the way selection was defined in the model (eliminating *aa* homozygotes). Second, the change in allelic frequency is proportional to q^2, which appears in the numerator of the expression. In other words, allelic frequency is declining as a relative function of the number of homozygous recessive individuals in the population. This fact is consistent with the premise of the selection model (with selection against the homozygous recessive genotype). This final formula supports the methodology of the model.

Equilibrium conditions

Next we calculate the equilibrium q by setting the Δq equation equal to zero, since a population in equilibrium will show no change in allelic frequencies from one generation to the next:

$$\frac{-q^2}{1+q} = 0 \quad \text{...(19)}$$

For a fraction to be zero, the numerator must equal zero. Thus, $q^2 = 0$, and $\hat{q} = 0$. At equilibrium, the *a* allele should be entirely removed from the population. If the *aa* homozygotes are being removed, and if there is no mutation to return *a* alleles to the population, then eventually the *a* allele disappears from the population.

Time frame for equilibrium

One shortcoming of this selection model is that it is not immediately apparent how many generations will be required to remove the *a* allele. The deficiency can be compensated for by using a computer simulation or by introducing a calculus differential into the model. This figure clearly shows that the *a* allele is removed more quickly when selection is stronger (when *s* is larger) and that the curves appear to be asymptotic—the *a* allele is not immediately eliminated and would not be entirely removed until an infinitely large number of generations had passed. There is a reason for the asymptotic behaviour of the graph: As the *a* allele becomes rarer and rarer, it tends to be found in heterozygotes. Since selection can remove only *aa* homozygotes, an *a* allele hidden in an *Aa* heterozygote will not be selected against. When $q = 0.5$, there are two heterozygotes for every *aa* homozygote. When $q = 0.001$, there are almost two thousand heterozygotes per *aa* homozygote. Remember, only the recessive homozygote is selected against. Natural selection cannot distinguish the dominant homozygote from the heterozygote.

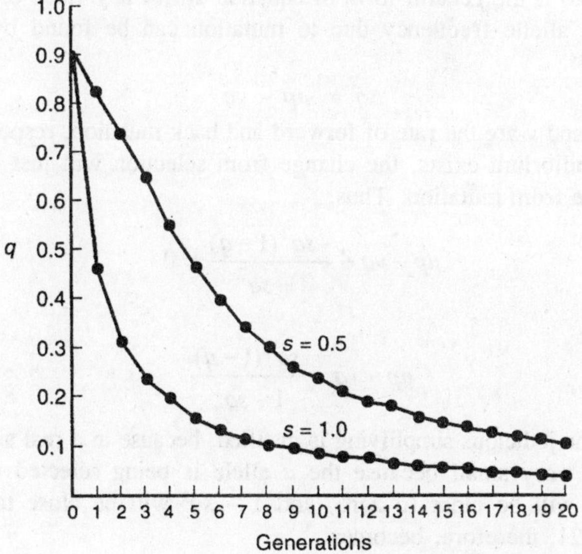

Fig. 7.4. Decline in q under different intensities of selection against the aa homozygote.

Selection-Mutation Equilibrium

Although a deleterious allele is eliminated slowly from a population, the time frame is so great that there is opportunity for mutation to bring the allele back. Given a population in which alleles are removed by selection and added by mutation, the point at which no change in allelic frequency occurs, the *selection-mutation equilibrium*, may be determined as follows. The new frequency (q_{n+1}) of the recessive *a* allele after nonlethal selection ($s < 1$) against the recessive homozygote is obtained by equation 15:

$$q_{n+1} = \frac{q(1-sq)}{1-sq^2}$$

Change in allelic frequency under this circumstance will thus be

$$\Delta q = q_{n+1} - q = \frac{q(1-sq)}{(1-sq^2)} - \frac{q(1-sq^2)}{(1-sq^2)}$$

$$= \frac{q - sq^2 - q + sq^3}{(1-sq^2)}$$

$$= \frac{-sq^2(1-q)}{1-sq^2} \qquad \text{...(20)}$$

Equation 20 is the general form of equation 18 for any value of *s*. The change in allelic frequency due to mutation can be found by using equation 4:

$$\Delta q = \mu p - vq$$

where μ, and v are the rate of forward and back mutation, respectively. When equilibrium exists, the change from selection will just balance the change from mutation. Thus,

$$\mu p - vq + \frac{-sq^2(1-q)}{1-sq^2} = 0$$

and

$$\mu p - vq = \frac{sq^2(1-q)}{1-sq^2} \qquad \text{...(21)}$$

Now, some judicious simplifying is justified, because in a real situation, *q* will be very small because the *a* allele is being selected against. Thus, *vq* will be close to zero, and $1 - sq^2$ will be close to unity. Equation 21, therefore, becomes:

$$\mu p \cong sq^2(1-q)$$

$$\mu(1-q) \cong sq^2(1-q)$$
$$q^2 \cong \mu / s$$
$$\hat{q} \cong \sqrt{\mu / s} \quad ...(22)$$

In the case of a recessive lethal, *s* would be unity, so

$$q^2 \cong \mu \text{ and } \hat{q} = \sqrt{\mu}$$

If a recessive homozygote has a fitness of 0.5 ($s = 0.5$) and a mutation rate, of 1×10^{-5}, the allelic frequency at selection-mutation equilibrium will be

$$q^2 \cong \sqrt{\mu / s} \cong \sqrt{1 \times 10^{-5} / 0.5} \cong \sqrt{2 \times 10^{-5}}$$
$$\cong 0.004$$

If the recessive phenotype were lethal, then

$$q^2 \cong \sqrt{\mu / s} \cong \sqrt{1 \times 10^{-5} / 1}$$
$$\cong 0.003$$

These are very low equilibrium values for the *a* allele.

Types of Selection Models

In view of the limited ways that fitnesses can be assigned, only a limited number of selection models are possible. All possible selection models if we assume that fitnesses are constants and the highest fitness is one. (You might now go through the list of models and determine the equilibrium conditions for each.) Note that two possible fitness distributions are missing. There is no model in which fitnesses are $1 - s$, 1, and 1 for the A_1A_1, A_1A_2, and A_2A_2 genotypes, respectively (remembering that $p = f[A_1]$ and $q = f[A_2]$). That model is for selection against the A_1A_1 homozygote. Some reflection should show that this is the same model as model 1, except that the A_1 allele is acting like a recessive allele. In other words, natural selection acts against A_1A_1 homozygotes, but not against the A_1A_2 and A_2A_2 genotypes. Thus, the model reduces to model 1 if we treat A_1 as the recessive allele and A_2 as the dominant allele. Similarly, the ($1 - s_1$, $1 - s_2$, 1) model is eliminated for the same reason (allele A_2 is acting like the dominant allele and A_1 like the recessive allele). We now describe the outcome of each of the models in the table.

In both models 1 and 3, selection is against genotypes containing the A_2 allele. Model 1, which we just derived in detail, is the model for a deleterious recessive allele. Almost any enzyme defect in a metabolic pathway fits this model, such as PKU, alkaptonuria, Tay-

Sachs disease, and so on. In model 3, however, natural selection can detect the heterozygote, as is the case with deleterious alleles that are not completely recessive. An example would be the hemoglobin anomaly called thalassemia, a disorder common in some European and Asian populations, that produces a severe anemia in homozygotes and a milder anemia in heterozygotes. It should be clear that selection can more quickly eliminate a partially recessive allele than a completely recessive allele because the allele can no longer "hide" in the heterozygote.

Dominant or semidominant alleles (model 3) are usually more quickly removed from a population because they are completely open to selection. It takes an infinite number of generations to remove a recessive lethal allele, but only one generation for natural selection to remove a completely dominant lethal allele (model 3, where $s_1 = s_2 = 1$). Examples of dominant deleterious traits in people are Huntington disease, facioscapular muscular dystrophy, and chondrodystrophy.

Model 2 is interesting because selection against the heterozygote leads to an unstable equilibrium at $q = 0.5$. If one heterozygote is removed by selection, one each of the two alleles is eliminated. However, if p and q are not equal (and thus not equal to 0.5), then one A_1 allele is not the same proportion of the A_1 alleles as one A_2 allele is of all the A_2 alleles. In other words, in a population of fifty individuals with $q = 0.1$ and $p = 0.9$, one A_2 allele is 10% (1/10) of the A_2 alleles, whereas one A_1 allele is only 1.1% (1/90) of the A_1 alleles. Removing one each of the two alleles causes a decrease in q. Therefore, a population following model 2 is at equilibrium at $p = q = 0.5$. However, this is an unstable equilibrium. Any perturbation that changes the allelic frequencies causes the rarer allele to be selected against and eventually removed from the population. An example is the maternal-fetal incompatibility at the Rh locus in human beings. The disease erythroblastosis occurs only in heterozygous fetuses ($Rh^+ Rh^-$) in Rh-negative ($Rh^- Rh^-$) mothers. Heterozygotes are, therefore, selected against.

In model 4, selection is against homozygotes. This model is called the *heterozygote advantage*, and we will derive its equilibrium condition because the results are important to evolutionary theory. At equilibrium

$$\Delta q = \frac{pq(s_1 p - s_2 q)}{\overline{W}} \qquad \ldots(23)$$

For this expression to be zero, either

$$p = 0,\ q = 0,\ \text{or}\ (s_1 p - s_2 q) = 0$$

If $p = 0$ or $q = 0$, the result is trivial; the equilibrium exists only because of the absence of one of the alleles. The more meaningful equilibrium occurs when $s_1p - s_2q = 0$. In that case

$$s_1p = s_2q \text{ or } s_1(1 - q) = s_2q$$

and

$$q^2 = \frac{s_1}{s_1 + s_2} \qquad \text{...(24)}$$

Since $p + q = 1$,

$$\hat{p} = \frac{s_2}{s_1 + s_2} \qquad \text{...(25)}$$

Several interesting conclusions follow. First, unlike the other models of selection, this model allows a population to maintain both alleles. We can demonstrate that this equilibrium is stable by graphing the Δq value against q. Such a graph appears, in which q is the frequency of allele A_2 and the fitnesses of genotypes A_1A_1, A_1A_2 and A_2A_2 are assumed to be 0.8, 1, and 0.7, respectively. Note that if the equilibrium is perturbed by an increase or decrease in q, the population returns to the point of equilibrium. Second, the equilibrium is independent of the original allelic frequencies since it involves only the selection coefficients, s_1 and s_2. Last, the equilibrium for each allele (equations 24 and 25) is directly proportional to the selection coefficient against the other allele. As the selection against A_1 increases (s_1 increases), the equilibrium shifts toward a higher value of q (more A_2 alleles).

8

HARDY-WEINBERG LAW

Population geneticists spend most of their time doing one of two things: describing the genetic structure of populations or theorizing on the evolutionary forces acting on populations. On a good day, these two activities mesh and true insights emerge. In this chapter, we will do all of the above. The first part of the chapter documents the nature of genetic variation at the molecular level, stressing the important point that the variation between individuals within a species is similar to that found between species. After a short terminologic digression, we begin the theory with the traditional starting point of population genetics, the Hardy-Weinberg law, which describes the consequences of random mating on allele and genotype frequencies. Finally, we see that the genotypes at a particular locus do fit the Hardy-Weinberg expectations and conclude that the population mates randomly.

No one knows the genetic structure of any species. Such knowledge would require a complete description of the genome and spatial location of every individual at one instant in time. In the next instant, the description would change as new individuals are born, others die, and most move, while their transmitted genes mutate and recombine. How, then, are we to proceed with a scientific investigation of evolutionary genetics when we cannot describe that which interests us the most? Population geneticists have achieved remarkable success by choosing to ignore the complexities of real populations and focusing on the evolution of one or a few loci at a time in a population that is assumed to mate at random or, if subdivided, to have a simple migration pattern. The success of this approach, which is seen in both theoretical and experimental investigations, has been impressive, as I hope the reader will agree by the end of this book. The approach is not without its detractors. Years ago, Ernst Mayr mocked this approach as "bean bag genetics." In so doing, he echoed a view held by many of the pioneers of our field that natural selection acts on highly interactive coadapted genomes whose evolution cannot be understood by considering the evolution of a few loci in isolation from all others. Although genomes are certainly coadapted, there is precious little evidence that there are strong interactions between most polymorphic alleles in natural populations. The modern view, spurred on by the rush of DNA sequence data, is that we can profitably study loci in isolation.

This chapter begins with a description of the genetic structure of the alcohol dehydrogenase locus, *ADH,* in *Drosophila. ADH* is but one locus in one species. Yet, its genetic structure is typical in most regards. Other loci in *Drosophila* and in other species may differ quantitatively, but not in their gross features.

DNA Variation in Drosophila

Although population genetics is concerned mainly with genetic variation within species, until recently only genetic variation with major morphological manifestations, such as visible, lethal, or chromosomal mutations, could be analyzed genetically. The bulk of genetically based variation was refractory to the most sensitive of experimental protocols. Variation was known to exist because of the uniformly high heritabilities of quantitative traits; there was simply no way to dissect it.

Today, all this has changed. With readily available polymerase chain reaction (PCR) kits, the appropriate primers, and a sequencing machine, even the uninitiated can soon obtain DNA sequences from several alleles in their favourite species. In fact, sequencing is so easy that data are accumulating more rapidly than they can be interpreted.

The 1983 paper "Nucleotide polymorphism at the alcohol dehydrogenase locus of *Drosophila melanogaster*," by Marty Kreitman, was a milestone in evolutionary genetics because it was the first to describe sequence variation in a sample of alleles obtained from nature. At the time, it represented a prodigious amount of work. Today, a mere 13 years later, an undergraduate could complete the study in a few weeks. The alcohol dehydrogenase locus in *D. melanogaster* has the typical exon-intron structure of eukaryotic genes. Only the 768 bases of the coding sequence are given along with its translation.

Kreitman sequenced 11 alleles from Florida (Fl), Washington (Wa), Africa (Af), Japan (Ja), and France (Fr). When the sequences were compared base by base, it turned out that they were not all the same. In fact, no two alleles had exactly the same DNA sequence, although within just the coding sequences.

Within the coding region of the 11 *ADH* alleles, 14 sites have two alternative nucleotides. A site with different nucleotides in independently sampled alleles is called a segregating site; less often, it is called a *polymorphic site*. About 1.8 of every 100 sites are segregating in the *ADH* sample, a figure that is typical for *D. melanogaster* loci. The variation at 13 of the 14 segregating sites is silent, so called because the alternative codons code for the same amino acid. The variation at the 578th nucleotide position results in a change of the amino acid at position 192 in the protein, where either a lysine (AAG) or a threonine (ACG) is found. A nucleotide polymorphism that causes an amino acid polymorphism is called a *replacement polymorphism*.

Table 8.1. The 11 ADH alleles.

Allele	39	226	387	393	441	513	519	531	540	578	606	615	645	684
Reference	T	C	C	C	C	C	T	C	C	A	C	T	A	G
Wa-S	.	T	T	.	A	A	C	.	.	.	.	.	.	.
F1-1S	.	T	T	.	A	A	C	.	.	.	.	.	.	.
Af-S	.	.	.	.	.	.	.	.	.	.	.	.	.	A
Fr-S	.	.	.	.	.	.	.	.	.	.	.	.	.	A
Fl-2S	G	.	.	.	.	.	.	.	.	.	.	.	.	.
Ja-S	G	.	.	.	.	.	.	.	T	.	T	.	C	A
Fl-F	G	.	.	.	.	.	.	G	T	C	T	C	C	.
Fr-F	G	.	.	.	.	.	.	G	T	C	T	C	C	.
Wa-F	G	.	.	.	.	.	.	G	T	C	T	C	C	.
Af-F	G	.	.	.	.	.	.	G	T	C	T	C	C	.
Ja-F	G	.	.	A	.	.	.	G	T	C	T	C	C	.

Kreitman's data pose a question which is the Great Obsession of population geneticists: What evolutionary forces could have led to such divergence between individuals within the same species? A related question that sheds light on the Great Obsession is: Why the preponderance of silent over replacement polymorphisms? The latter question is more compelling when you consider that about three-quarters of random changes in a typical DNA sequence will cause an amino acid change. Rather than 75 percent of the segregating sites being replacement, only 7 percent are replacement. Perhaps silent variation is more common because it has a very small effect on the phenotype. By contrast, a change in a protein could radically alter its function.

Alcohol dehydrogenase is an important enzyme because flies and their larvae are often found in fermenting fruits with a high alcohol concentration. Inasmuch as alcohol dehydrogenase plays a role in the detoxification of ingested alcohol, a small change in the protein could have substantial physiologic consequences. Thus, it is reasonable to suggest that selection on amino acid variation in proteins will be stronger than on silent variation and that the stronger selection might reduce the level of polymorphism. This is a good suggestion, but it is only a suggestion. Population geneticists take such suggestions and turn them into testable scientific hypotheses, as will be seen as this book unfolds.

Just as there is *ADH* variation within species, so too is there variation between species. In this figure, the coding region of the *ADH* locus in *D. melanogaster* is compared to that of the closely related species, *D. erecta*. Thirty-six of 768 nucleotides differ between the two species. The probability that a randomly chosen site is different is 36/768 = 0.0468; note that this is also the average number of nucleotide differences per site. Of the 36 differences, only 10 (26%) result in amino acid differences between the two species. Kreitman's polymorphism data also exhibited less replacement than silent variation, but the disparity was somewhat greater: one replacement difference out of 14 (7%) segregating sites.

The comparison of variation within and between species shows no striking lack of congruence. In both cases, all of the differences involve only isolated nucleotides and, in both cases, there are more silent than replacement changes. Things could have been otherwise. For example, the variation within species could have involved isolated nucleotide changes while the differences between species could have been due to insertions and deletions. Were this observed, then the variation within species would have little to contribute to our understanding of evolution in the broader sense. As it is, population geneticists feel confident that their studies of variation within populations play a key role in the wider discipline of evolutionary biology.

Molecular variation may seem far removed from what interests most evolutionists. For many, the allure of evolution is the understanding of the processes leading to the strange creatures of the past or the sublime adaptations of modern species. The raw material of this evolution, however, is just the sort of molecular variation described above. This genetically determined variation must ultimately be due to the kind of molecular variation observed at the *ADH* locus. As of this writing, the connections between molecular variation and phenotypic variation have not been made. The discovery of these connections remains one of the great frontiers of population genetics. Of particular interest in this endeavor will be the relative roles played by variation in coding regions, as seen in the *ADH* example, and variation in the control regions just upstream from coding regions.

Loci and Alleles

We must now make a short digression into vocabulary because two words, *locus* and *allele*, must be made more precise than is usual in genetics textbooks. Although the terms were used without ambiguity for many years, the increase in our understanding of molecular genetics has clouded their original meanings considerably. Here we will use *locus* to refer to the place on a chromosome where an allele resides. An *allele* is just the bit of DNA at that place. A locus is a template for an allele. An allele is an instantiation of a locus. A locus is not a tangible thing; rather, it is a map describing where to find a tangible thing, an allele, on a chromosome. (Some books use *gene* as a synonym for our *allele*. However, *gene* has been used in so many different contexts that it is not very useful for our purposes.) With this convention, a diploid individual may be said to have two alleles at a particular autosomal locus, one from its mother and the other from its father.

Population genetics, like other areas of genetics, is concerned with alleles that differ one from another. However, in population genetics there are subtleties in what is meant by "different alleles." There are three fundamental ways in which alleles at the same locus may differ:

By origin. Alleles differ by origin if they come from the same locus on different chromosomes. One often refers to a sample of n (different) alleles from a population. What is meant by "different" in this context is "different by origin." For example, the two alleles at a specified locus in a diploid individual are always different by origin. The 11 alleles in Kreitman's sample also differ by origin.

By state. Whether or not two alleles are said to differ by state depends on the context. If the context is the DNA sequence of the alleles, then they are different by state if they have different DNA sequences. The difference may as small as one nucleotide out of thousands. However, in evolutionary studies we frequently focus on particular aspects of alleles and may choose to put them in different states depending on the nature of the difference. For example, if our interest is in protein evolution, we may choose to say that two alleles are different by state if and only if they differ in their amino acid sequences. (We do this in full recognition that some alleles with the same amino acid sequence may have different DNA sequences as a consequence of the redundancy of the genetic code.) Similarly, we may choose to call two alleles different by state if and only if they have different amino acids at a particular site, perhaps at the fourth position in the protein. States may also be thought of *as* phenotypes, which could include the DNA sequence, the protein sequence, the colour of the pea, or other genetically determined phenotypes of interest.

By descent. Alleles differ by descent when they do not share a common ancestor allele. Strictly speaking, two alleles from the same locus can never be different by descent as all contemporary alleles share a remote common ancestor. In practice, we are often concerned with a relatively short time in the past and are content to say that two alleles are different by descent if they do not share a common ancestor allele in, say, the past 10 generations. Two alleles that are different by descent may or may not be different by state because of mutation.

The converse of the above involves identity by origin, state, or descent. Alleles that are identical by origin are necessarily identical by state and descent. Two alleles that are identical by descent may not be identical by state because of mutation. A simple example of three nucleotides in alleles obtained from two individuals in generation n and traced back to their ancestor allele in generation $n - 2$. The two alleles are identical by descent because they are both copies of the same ancestor allele in the recent past. However, they are different by state because a mutation from c to g appeared in the right-hand allele.

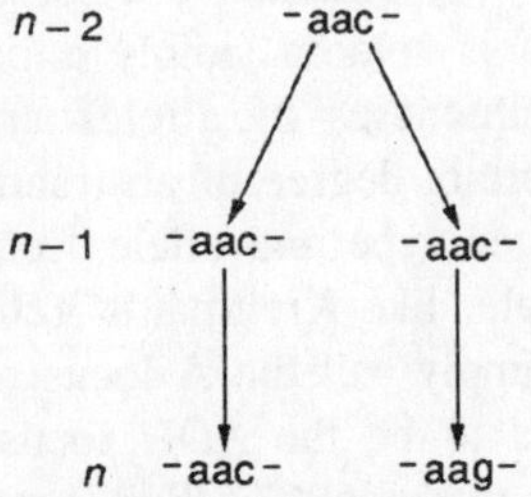

Fig. 8.1. Two alleles in generation n that are identical by descent but differ in state.

Diploid individuals are said to be heterozygous at a locus if the two alleles at that locus are different by state. They are homozygous if their two alleles are identical by state. The use of homozygous or heterozygous is always in the context of the states under study. If we are studying proteins, we may call an individual homozygous at a locus when the protein sequences of the two alleles are identical, even if their DNA sequences are different.

Originally, *alleles* referred to different states of a gene. Our definition differs from this traditional usage in that alleles exist even if there is no genetic variation at a locus. Difference by origin has not been used before. It is introduced here to be able to use phrases like "a sample of *n* different alleles" without implying that the alleles are different by state.

Kreitman's sample contains 11 alleles that differ by origin. How many alleles differ by state? If we were interested in the full DNA sequence, then the sample contains six alleles that are different by state. If we were interested in proteins, then the sample contains only two alleles that differ by state. Of the two protein alleles, the one with a lysine at position 192 makes up $6/11 = 0.55$ of the alleles. The usual way to say this is that the allele frequency of the lysine-containing allele in the sample is 0.55. The sample allele frequency is an estimate of the population allele frequency. It's not a particularly precise estimate because of the small sample size. A rough approximation to the 95 percent confidence interval for a proportion is

$$\hat{p} \pm 1.96\sqrt{\hat{p}(1-\hat{p})/n},$$

where $\hat{p}$ is the estimate of the proportion, 0.55 in our case, and n is the sample size. Thus, the probability that the population allele frequency falls within the interval (0.26, 0.84) is 0.95. If a more precise estimate is needed, the sample size would have to be increased.

Genotype and Allele Frequencies

Population genetics is very quantitative. A description of the genetic structure of a population is seldom simply a list of genotypes, but rather uses relative frequencies of alleles and genotypes. With quantification comes a certain degree of abstraction. For example, to introduce the notion of genotype and allele frequencies we will not refer to a particular sample, like Kreitman's *ADH* sample, but rather to a locus that we will simply call the A locus. (No harm will come in imagining the *A* locus to be the *ADH* locus.) Initially, we will assume that the locus has two alleles, called A_1 and A_2, segregating in the population. (These could be the two protein alleles at the *ADH* locus.) By implication, these two alleles are different by state. There will be three genotypes in the population: two homozygous genotypes,

$A_1 A_1$ and A_2A_2, and one heterozygous genotype, $A_1 A_2$. The relative frequency of a genotype will be written x_{ij}, as illustrated in the following table.

Genotype:	$A_1 A_1$	A_1A_2	A_2A_2
Relative frequency:	x_{11}.	x_{12}	x_{22}

As the relative frequencies must add to one, we have

$$x_{11} + x_{12} + x_{22} = 1.$$

The ordering of the subscripts for heterozygotes is arbitrary. We could have used x_{21} instead of x_{12}. However, it is not permissible to use both. In this book, we will always use the convention of making the left index the numerically smaller one.

Allele frequencies play as important a role in population genetics as do genotype frequencies. The frequency of the A_1 allele in the population is

$$p = x_{11} + \frac{1}{2}x_{12}, \qquad \text{...1}$$

and the frequency of the A_2 allele is

$$q = 1 - p = x_{22} + \frac{1}{2}x_{12}.$$

We can think of the allele frequency, *p,* in two different ways. One is simply as the relative frequency of A_1 alleles among all of the *A* alleles in the population. The other is as the probability that an allele picked at random from the population is an A_1 allele. The act of picking an allele at random may be broken down into a sequence of two actions: picking a genotype at random from the population and then picking an allele at random from the chosen genotype. Because there are three genotypes, we could write *p* as

$$p = (x_{11} \times 1) + (x_{12} \times \frac{1}{2}) + (x_{22} \times 0).$$

This representation shows that there are three mutually exclusive ways in which we might obtain an A_1 allele and gives the probability of each. For example, the first term in the sum is the joint event that an A_1A_1 is chosen (this occurs with probability x_{11}) and that an A_1 allele is subsequently chosen from the A_1A_1 individual (this occurs with probability one). It is difficult to underestimate the importance of probabilistic reasoning when doing population genetics. I urge the reader to think carefully about the probabilistic definition of *p* until it becomes second nature.

Most loci have more than two alleles. In such cases, the frequency of the *i*th allele will be called p_i. As before, the frequency of the A_iA_j genotype will be called x_{ij}. For heterozygotes, $i \neq j$ and, by convention, $i < j$. As with the two-allele case, the sum of all of the genotype frequencies must add to one. For example, if there are *n* alleles, then

$$1 = x_{11} + x_{22} + \cdots + x_{nn} + x_{12} + x_{13} + \cdots + x_{(n-1)n}$$
$$= \sum_{i=1}^{n} \sum_{j \geq i}^{n} x_{ij}.$$

The frequency of the ith allele is

$$p_i = x_{ii} + \frac{1}{2}\sum_{j=1}^{i-1} x_{ji} + \frac{1}{2}\sum_{j=i+1}^{n} x_{ij}$$

Again, this allele frequency has both a relative frequency and a probabilistic interpretation.

In the mid-1960s, population geneticists began to use electrophoresis to describe genetic variation in proteins. For the first time, the genetic variation at a "typical" locus could be ascertained. Harry Harris's 1966 paper, "Enzyme polymorphism in man," was among the first of many electrophoretic survey papers. In it, he summarized the electrophoretic variation at 10 loci sampled from the English population. The protein produced by one of these loci is placental alkaline phosphatase. Harris found three phosphatase alleles that differed by state (migration speed) and called them S (slow), I (intermediate), and F (fast) for their rate of movement in the electrophoresis apparatus.

Table 8.2. The frequencies of alkaline phosphatase genotypes in a sample from the people.

Genotype	*Number*	*Frequency*	*Expected*
SS	141	0.4247	0.4096
SF	111	0.3343	0.3507
FF	28	0.0843	0.0751
SI	32	0.0964	0.1101
FI	15	0.0452	0.0471
II	5	0.0151	0.0074
Total	332	1.0000	1.0000

The frequency of heterozygotes at the placental alkaline phosphatase locus is 158/332 = 0.48, which is unusually high for human protein loci. The average probability that an individual is heterozygote at a locus examined in this paper is approximately 0.05. If this could be extrapolated to the entire genome, then a typical individual would be heterozygous at 1 (at least) of every 20 loci. However, there is evidence that the enzymes used in Harris's study are not "typical" loci. They appear to be more variable than other protein loci. At present, we do not have a reliable estimate of the distribution of protein heterozygosities across loci for any species.

Randomly Mating Populations

The first milestone in theoretical population genetics, the celebrated Hardy-Weinberg law, was the discovery of a simple relationship between allele frequencies and genotype frequencies at an autosomal locus in an equilibrium randomly mating population. For example, the S allele is more frequent than the F allele and the SS homozygote is more frequent than the FF homozygote, suggesting that homozygotes of more frequent alleles will be more common than homozygotes of less frequent alleles. Such qualitative observations yield quite naturally to the desire for quantitative relationships between allele and genotype frequencies, as provided by the insights of George Hardy and Wilhelm Weinberg.

The Hardy-Weinberg law describes the equilibrium state of a single locus in a randomly mating diploid population that is free of other evolutionary forces, such as mutation, migration, and genetic drift. By random mating, we mean that mates are chosen with complete ignorance of their genotype (at the locus under consideration), degree of relationship, or geographic locality. For example, a population in which individuals prefer to mate with cousins is not a randomly mating population. Rather, it is an inbreeding population. A population in which $A_1 A_1$ individuals prefer to mate with other A_1A_1 individuals is not a randomly mating population either. Rather, this population is experiencing assortative mating. Geography can also prevent random mating if individuals are more likely to mate with neighbours than with mates chosen at random from the entire species. Assortative mating will not be discussed further because it is a specialized topic, although one that can play an important role in the evolution of some species.

The Hardy-Weinberg law is particularly easy to understand in hermaphroditic species (species in which each individual is both male and female). The autosomal loci of hermaphrodites reach their Hardy-Weinberg equilibrium in a single generation of random mating, no matter how far the initial genotype frequencies are from their equilibrium values. Our task, then, is to study the change in genotype frequencies in hermaphrodites brought about by random mating at an autosomal locus with two alleles, A_1 and A_2, and genotype frequencies x_{11}, x_{12}, and x_{22}.

To form a zygote in the offspring generation, the assumption of random mating requires that we choose two gametes at random from the parent generation. The probability that the zygote is an A_1A_1 homozygote is the product of the probability that the egg is A_1, p, times the probability that the sperm is A_1, also p. (The fact that these two probabilities are the same is the consequence of assuming that the species is hermaphroditic.) Thus, the probability that a randomly formed zygote is A_1A_1 is just p^2 by the product rule of probabilities for independent events. Similarly, the probability that a randomly formed zygote is A_2A_2 is q^2. An A_1A_2 heterozygote may be formed in two different ways. One way is with an A_1 egg and an A_2 sperm. The probability of this combination is pq. The other way is with an A_2 egg and an A_1 sperm. The probability of this combination is also pq. Thus, the total probability of forming a heterozygote is $2pq$ by the addition rule of probabilities for mutually exclusive events.

After one round of random mating, the frequencies of the three genotypes are

Genotype:	A_1A_1	A_1A_2	A_2A_2
Frequency (H-W):	p^2	$2pq$	q^2

These are the Hardy-Weinberg genotype frequencies. As advertised, they depend only on the allele frequencies: If you know p, then you know the frequencies of all three genotypes.

The important things to note about the evolutionary change brought about by random mating in diploid hermaphroditic populations are:

1. The frequencies of the alleles do not change as a result of random mating, as may be seen by using Equation 1 with the Hardy-Weinberg frequencies. Random mating can change genotype frequencies, not allele frequencies. Consequently, the Hardy-Weinberg genotype frequencies will remain unchanged in all generations after the first.
2. The Hardy-Weinberg equilibrium is attained in only one round of random mating. This is traceable to our assumption that the species is hermaphroditic (and that we are studying an autosomal locus). In a species with separate sexes, it takes two generations to achieve Hardy-Weinberg equilibrium, as we will soon discover.

3. To calculate the genotype frequencies after a round of random mating, we need only the allele frequencies before random mating, not the genotype frequencies.

Of course, many species are not hermaphrodites but are dioecious; each individual is either male or female. To further complicate matters, the genotype frequencies could be different in the two sexes. As an extreme example, suppose that all of the females are A_1A_1 and all of the males are A_2A_2. If the sexes are equally frequent, the frequency of the A_1 allele in the population is $p = 1/2$. After one round of random mating, the frequencies of the A_1A_1 and A_2A_2 homozygotes are zero, and the frequency of the A_1A_2 heterozygote is one. These frequencies are far from the Hardy-Weinberg frequencies. However, the third generation, produced by random mating of heterozygotes, has genotypes A_1A_1, A_1A_2, and A_2A_2 in the Hardy-Weinberg frequencies 1/4, 1/2, and 1/4, respectively. Thus, for dioecious species with unequal genotype frequencies in the two sexes, it can take two generations to reach equilibrium.

Can it take more or fewer? The answer depends on whether the locus is on an autosome or a sex chromosome. For now, consider only the case of an autosomal locus, for which one round of random mating makes the allele frequencies the same in both sexes and equal to the average of the frequencies in the males and females of the parent or first generation. Call the frequency of the A_1 allele in the first and second generations p.

In the next generation (the third), the probability that a zygote is A_1A_1 is the product of the probabilities that the sperm is A_1, p, and that the egg is A_1, also p. These two probabilities became equal in the second generation. From here on, the argument parallels that used for hermaphrodites with the same ultimate genotype frequencies. Thus, if the allele frequencies are different in the two sexes, it takes two generations to reach Hardy-Weinberg frequencies. Otherwise, it takes only one generation.

One of the most important consequences of the Hardy-Weinberg law concerns the genotypes occupied by rare alleles. Suppose the A_2 allele is rare; that is, $q = 1 - p$ is small. Are A_2 alleles more likely to be in A_2A_2 homozygotes or A_1A_2 heterozygotes? The ratio of the latter to the former is

$$\frac{2pq}{q^2} = \frac{2p}{q} \approx \frac{2}{q}.$$

The approximation used in the last step makes use of the assumption that q is small. As $p = 1 - q$, p may be approximated by one because q is small relative to one. For example, if q is about 0.01, the error in this approximation is about 1 percent, which is perfectly acceptable for population genetics. If $q = 0.01$, an A_2 allele is about 200 times more likely to be in a heterozygote than in a homozygote. If $q = 0.001$, it is about 2000 times more likely to be in a heterozygote. Thus, rare alleles mostly find themselves in heterozygotes and, as a consequence, their fate is tied to their dominance relationship with the A_1 allele. This is our first clue that dominance is an important factor in evolution.

The generalization of the Hardy-Weinberg law to multiple alleles requires no new ideas. Let the frequency of the k alleles, A_i, $i = 1 \ldots k$, be p_i, $i = 1 \ldots k$. Using the same argument as before, it should be clear that the frequency of the A_iA_i homozygote after random mating will be p_i^2 and the frequency of the A_iA_j heterozygote will be $2p_ip_j$. The total frequency of homozygotes is given by

$$G = \sum_{i=1}^{k} p_i^2. \qquad \ldots(2)$$

G is called the homozygosity of the locus. The heterozygosity of the locus is given by

$$H = 1 - G = 1 - \sum_{i=1}^{k} p_i^2.$$

For randomly mating diploid populations, the heterozygosity equals the frequency of heterozygotes. Note, however, that the definition of heterozygosity uses only allele frequencies, not genotype frequencies. Because of this, heterozygosity is often used to describe levels of variation in populations that do not conform to the Hardy-Weinberg assumption of random mating. It is even used to describe variation *in* bacterial populations, which are haploids.

The frequencies of the S, F, and I alleles of placental alkaline phosphatase are 0.640, 0.274, and 0.086, respectively. From these we can calculate the expected frequencies of each genotype in the population assuming that it is in Hardy-Weinberg equilibrium. The expected frequency of SS homozygotes, for example, is $0.64^2 = 0.4095$. The agreement between the observed and expected numbers is quite good. A χ^2 test does not allow rejection of the Hardy-Weinberg hypothesis at the 5 percent level. The human placental alkaline phosphatase story is typical of many proteins examined from populations that are thought to mate at random. Very few cases of significant departures from Hardy-Weinberg expectations have been recorded in outbreeding species.

A description of the genetic structure of a population must include a geographic component if the ultimate goal is to understand the evolutionary forces responsible for genetic variation. A conjecture about the evolutionary history of the alkaline phosphatase alleles, for example, will be of one sort if the allele frequencies are the same in all subpopulations and of quite another sort if the subpopulations vary in their allele frequencies. Some representative frequencies taken from the 1988 compilation of human polymorphism data by Arun Roy choudhury and Masatoshi Nei. As is apparent, there is considerable geographic variation in the frequencies of the three alleles. We must conclude what will be obvious to most: The human population is not one large, randomly mating population. The agreement of the genotype frequencies with Hardy-Weinberg expectations within the English population, however, suggests that local groupings have historically approximated randomly mating populations. Most other species show a similar pattern. Some have less differentiation between geographic areas, others quite a bit more. But most show some differentiation, and this fact should be incorporated into our view of the genetic structure of populations.

Table 8.3. Geographic variation in the three alleles of placental alkaline phosphatase in human.

Place	*S*	*F*	*I*	*Others*	*n*
England	.637	.270	.085	.008	597
Italy	.661	.256	.075	.007	273
West India	.701	.217	.066	.016	208
Thailand	.746	.081	.165	.008	188
Japan	.724	.038	.236	.003	294
Nigeria	.942	.019	.039		130
Canadian Inuits	.556	.142	.296	.006	81
Papua New Guinea	.880	.050	.068	.002	338

A great deal of work has been done to estimate the levels of genetic variation in natural populations. For electrophoretically detectable protein variation, average heterozygosities (averaged across loci) vary from zero to about 0.15. The average protein heterozygosity in humans is about 0.05; for *D. melanogaster* it is about 0.12. Average heterozygosities are somewhat misleading because they bury the fact that there is a tremendous amount of variation between loci in levels of polymorphism. For example, soluble enzymes are much more polymorphic than are abundant nonenzymatic proteins. Nonetheless, average heterozygosities do give the correct impression that there is a lot of genetic variation in natural populations.

Why does the Hardy-Weinberg law play such a central role in population genetics? Consider that life first appeared on Earth about four billion years ago. For the next two billion or so years, the earth was populated by haploid prokaryotes, not diploid eukaryotes. During this time, most of the basic elements of living systems appeared: the genetic code, enzymes, biochemical pathways, photosynthesis, bipolar membranes, structural proteins, and on and on. Thus, the most fundamental innovations evolved in populations where the Hardy-Weinberg law is irrelevant! If population genetics were primarily concerned with the genetic basis of evolution, then it is odd that the Hardy-Weinberg law is introduced so early in most texts (including this one). One might expect to find a development targeted at those important first two billion years with a coda to handle the diploid upstarts.

Like so much of science, the development of population genetics is anthrocentric. Our most consuming interest is with ourselves. In fact, the Weinberg of Hardy-Weinberg was a human geneticist struggling with the study of inheritance in a species in which setting up informative crosses is frowned upon. *Drosophila* ranks a close second to *Homo* in the eyes of geneticists. Both humans and fruit flies exhibit genetic variation in natural populations, and this variation demands an evolutionary investigation. Population genetics and its Great Obsession grew out of this fascination with variation in species we love, not out of a desire to explain the origin of major evolutionary novelties.

9

GENETIC DRIFT

The discussion of random mating and the Hardy-Weinberg law in the previous chapter was premised on the population size being infinite. Sometimes real populations are very large (roughly 10^9 for our own species), in which case the infinite assumption might seem reasonable, at least as a first approximation. However, the population sizes of many species are not very large. Bird watchers will tell you, for example, that there are fewer than 100 Bachman's warblers in the cypress swamps of South Carolina. For these warblers, the infinite population size assumption of the Hardy-Weinberg law may be hard to accept. In finite populations, random changes in allele frequencies result from variation in the number of offspring between individuals and, if the species is diploid and sexual, from Mendel's law of segregation.

Genetic drift, the name given to these random changes, affects evolution in two important ways. One is as a dispersive force that removes genetic variation from populations. The rate of removal is inversely proportional to the population size, so genetic drift is a very weak dispersive force in most natural populations. The other is drift's effect on the probability of survival of new mutations, an effect that is important even in the largest of populations. In fact, we will see that the survival probability of beneficial mutations is (approximately) independent of the population size.

The dispersive aspect of genetic drift is countered by mutation, which puts variation back into populations. We will show how these two forces reach an equilibrium and how they can account for much of the molecular variation described already. The neutral theory states that much of molecular variation is due to the interaction of drift and mutation. This theory, one of the great accomplishments of population genetics because it is the first fully developed theory to satisfy the Great Obsession, has remained controversial partly because it has been difficult to test and partly because of its seemingly outrageous claim that most of evolution is due to genetic drift rather than natural selection, as Darwin imagined. The theory will be developed in this chapter and will reappear in several later chapters as we master additional topics relevant to the theory.

Computer Simulation

Simple computer simulations may be used to illustrate the consequences of genetic drift. These particular simulations model a population of $N = 20$ diploid individuals with two segregating alleles, A_1 and A_2. The frequency of A_1 at the start of each simulation is $p = 0.2$, which represents 8 A_1 alleles and 32 A_2 alleles. Each new generation is obtained from the previous generation by repeating the following three steps $2N = 40$ times.

1. Choose an allele at random from among the $2N$ alleles in the parent generation.
2. Make an exact copy of the allele.
3. Place the copy in the new generation.

After 40 cycles through the algorithm, a new population is created with an allele frequency that will, in general, be different from that of the original population. The reason for the difference is the randomness introduced in step 1.

As written, these steps may be simulated on a computer or with a bag of marbles of two colours, initially 8 of one colour and 32 of another (providing that you have all of your marbles). Obviously, allele frequencies do change at random. Nothing could be farther from the constancy promised by Hardy-Weinberg.

In natural populations, there are two main sources of randomness. One is Mendel's law of segregation. When a parent produces a gamete, each of its two homologous alleles is equally likely to appear in the gamete. The second is demographic stochasticity.* Different individuals have different numbers of offspring for complex reasons that collectively appear to be random. Neither of these sources gives any preference to particular alleles. Each of the $2N$ alleles in the parent generation has an equal chance of having a copy appear in the offspring generation.

You may have noticed that the computer simulations do not explicitly incorporate either segregation or demographic stochasticity, even though these two sources of randomness are the causes of genetic drift. Nonetheless, they do represent genetic drift as conceived by most population geneticists. A more realistic simulation with both sources of randomness would behave almost exactly like our simple one. Why then, do we use the simpler simulation? The answer is a recurring one in population genetics: The simpler model is easier to understand, is easier to analyze mathematically, and captures the essence of the biological situation. With drift, the essence is that each allele in the parental generation is equally likely to appear in the offspring generation. In addition, the probability that a particular allele appears in the offspring generation is nearly independent of the identities of other alleles in the offspring generation. The simple algorithm in the simulation does have both of these properties and simulates what is called the Wright-Fisher model in honor of Sewall Wright and R. A. Fisher, two pioneers of population genetics who were among the first to investigate genetic drift.

One, of course, is that genetic drift causes random changes in allele frequencies. Each of the five populations behaves differently even though they all have the same initial allele frequency and the same population size. By implication, evolution can never be repeated. A second feature is that alleles are lost from the population. In two cases, the A_1 allele was lost; in two other cases, the A_2 allele was lost. In the fifth case, both alleles are still in the population after 100 generations. From this we might reasonably conclude that genetic drift removes genetic variation from populations. The third feature is more subtle: the direction of the random changes is neutral. There is no

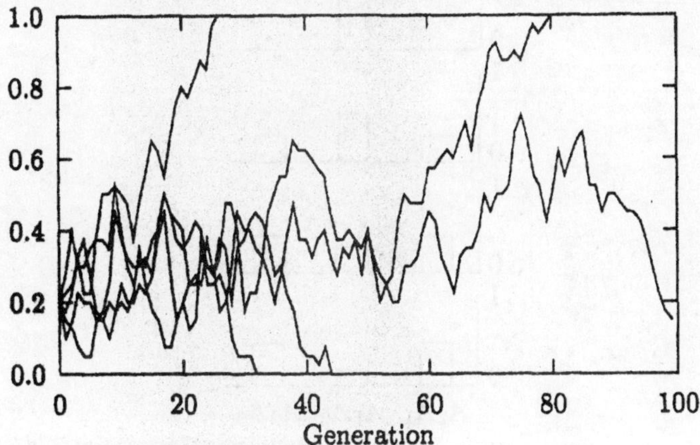

Fig. 9.1. A computer simulation of genetic drift.

systematic tendency for the frequency of alleles to move up or down. A few simulations cannot establish this feature with certainty.

Decay of Heterozygosity

As a warm-up to our general treatment of genetic drift, we will first examine the simplest non-trivial example of genetic drift: a population made up of a single hermaphroditic individual. If the individual is an A_1A_2 heterozygote, the frequency of the A_1 allele in the population is one-half. When the population reproduces by mating at random (a strange notion, but accurate) and the size of the population is kept constant at one, the heterozygote is replaced by an A_1A_1, A_1A_2, or A_2A_2 individual with probabilities 1/4, 1/2, and 1/4, respectively. These probabilities are the probabilities that the allele frequency becomes 1, 1/2, or 0 after a single round of random mating. In the first or third outcome, the population is a single homozygote individual and will remain homozygous forever. In the second outcome, the composition of the population remains unchanged.

After another round of random mating, the probability that the population is a heterozygous individual is 1/4, which is the probability that it is heterozygous in the second generation, 1/2, times the probability that it is heterozygous in the third generation given that it is heterozygous in the second generation, 1/2. It should be clear that the probability that the population is a heterozygote after t generations of random mating is $(1/2)^t$, which approaches zero as t increases. On average, it takes only two generations for the population to become

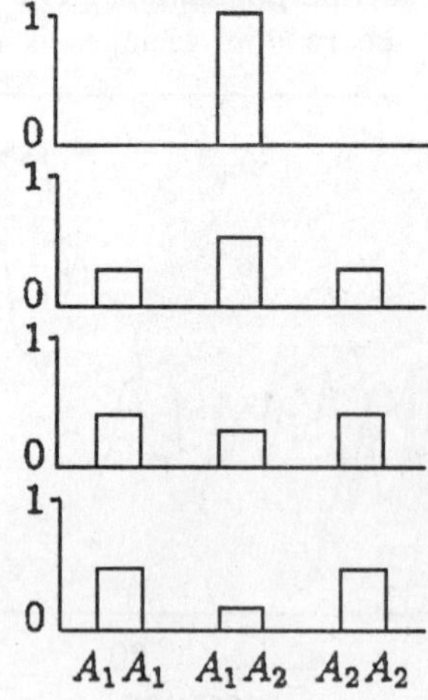

Fig. 9.2. Genotype frequencies for four generations of drift with N = 1.

homozygous. When it does, it is as likely to be homozygous for the A_1 allele as for the A_2 allele.

While simple, this example suggests the following features about genetic drift, some of which overlap our observations.

1. Genetic drift is a random process. The outcome of genetic drift cannot be stated with certainty. Rather, either probabilities must be assigned to different outcomes or the average outcome must be described.
2. Genetic drift removes genetic variation from the population. The probability that an individual chosen at random from the population is heterozygous after t generations of random mating is

$$H_t = H_0\left(1-\frac{1}{2N}\right)^t,$$

where H_0 is the initial probability of being a heterozygote (one in our example) and N is the population size (one in our example). This provocative form for our simple observation that $H_t = (1/2)^t$ adumbrates the main result of this section.
3. The probability that the ultimate frequency of the A_1 allele is one is equal to the frequency of the A_1 allele in the starting population, one-half.

The mathematical description of genetic drift can be quite complicated for populations with more than one individual. Fortunately, there is a simple and elegant way to study one of the most important aspects of genetic drift: the rate of decay of heterozygosity. As usual, we will be studying an autosomal locus in a randomly mating population made up of N diploid hermaphroditic individuals. The state of the population will be described by the variable G, defined to be the probability that two alleles different by origin (equivalently, drawn at random from the population without replacement) are identical by state. These alleles are assumed to be completely equivalent in function and, thus, equally fit in the eyes of natural selection. Such alleles are called neutral alleles. G is a measure of the genetic variation in the population, which is almost the same as the homozygosity of the population. When there is no genetic variation, $G = 1$. When every allele is different by state from every other allele, $G = 0$.

The value of G after one round of random mating, G', as a function of its current value, is

$$G' = \frac{1}{2N} + \left(1-\frac{1}{2N}\right)G. \quad ...1$$

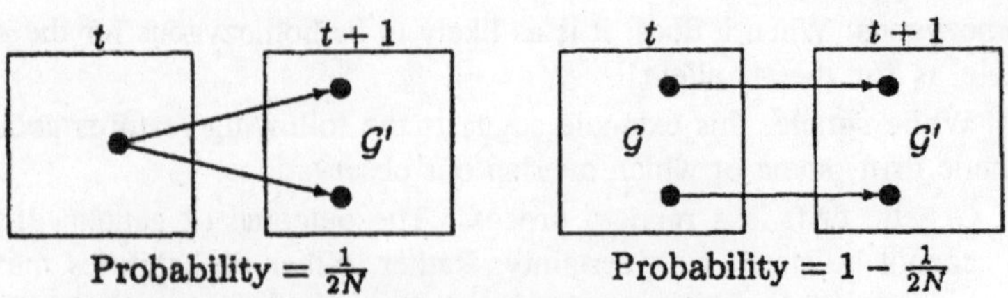

Fig. 9.3. The derivation of G'. The circles represents alleles, the arrows indicate the ancestry of the alleles.

The derivation, goes as follows. G' is the probability that two alleles that are different by origin in the next generation, called generation $t + 1$, are identical by state. Identity by state could happen in two different ways. One way is when the two alleles are copies of the same allele in the previous generation. The probability that the two alleles do share an ancestor allele in the previous generation is $1/(2N)$. (Pick one allele, and the probability that the allele picked next has the same parent allele as the first is $1/(2N)$, as all alleles are equally likely to be chosen.) The second way for the alleles to be identical by state is when the two alleles do not have the same ancestor allele in the previous generation, but their two ancestor alleles are themselves identical by state. This ancestry occurs with probability $1-1/(2N)$, and the probability that the two ancestor alleles are identical by state is G (by definition). As these two events are independent, the probability of the second way is $[1-1/(2N)]G$. Finally, as the two ways to be identical by state are mutually exclusive, the full probability of identity by state in the next generation is obtained by summation.

The time course for G is most easily studied by using

$$H = 1 - G,$$

the probability that two randomly drawn alleles are different by state. (H is similar to the heterozygosity of the population.) From Equation 1 and a few algebraic manipulations, we have

$$H' = 1 - G' = \left(1 - \frac{1}{2N}\right)H,$$

and finally,

$$\Delta_N H = -\frac{1}{2N}H \qquad \text{...(2)}$$

The Δ operator is used to indicate the change in a state variable that occurs in a single generation, $\Delta_N H = H' - H$. The subscript N in $\Delta_N H$ is simply a reminder that the change is due to genetic drift.

Equation 2 shows that the probability that two alleles are different by state decreases at a rate $1/(2N)$ each generation. For very large populations, this decrease will be very slow. Nonetheless, the eventual result is that all of the variation is driven from the population by genetic drift.

The full time course for H is

$$H_t = H_0\left(1-\frac{1}{2N}\right)^t, \qquad \ldots(3)$$

where H_t is H in the tth generation. The easiest way to show this is to examine the first few generations,

$$H_1 = H_0\left(1-\frac{1}{2N}\right)$$

$$H_2 = H_1\left(1-\frac{1}{2N}\right),$$

$$= H_0\left(1-\frac{1}{2N}\right)^2,$$

and then make a modest inductive leap to the final result.

Equation 3 shows that the decay of H is geometric. The probability that two alleles are different by state goes steadily down but does not hit zero in a finite number of generations. Nonetheless, the probability eventually becomes so small that most populations will, in fact, be homozygous. Every allele will be a descendent of a single allele in the founding population. All but one of the possibly thousands or millions of alleles in any particular population will fail to leave any descendents.

For large populations, genetic drift is a very weak evolutionary force, as may be shown by the number of generations required to reduce H by one-half. This number is the value of t that satisfies the equation $H_t = H_0/2$,

$$\frac{H_0}{2} = H_0\left(1-\frac{1}{2N}\right)^t.$$

Cancel H_0 from both sides, take the natural logarithm of both sides, and solve for t to obtain

$$t_{1/2} = \frac{-\ln(2)}{\ln(1-1/2N)}. \qquad \ldots(4)$$

The approximation of the log given

$$\ln(1+x) \approx x,$$

allows us to write

$$t_{1/2} \approx 2N \ln(2).$$

In words, the time required for genetic drift to reduce H by one-half is proportional to the population size.

When studying population genetics, placing results in a more general context is often enlightening. For example, a population of one million individuals requires about 1.38×10^6 generations to reduce H by one-half. If the generation time of the species were 20 years, it would take about 28 million years to halve the genetic variation. In geologic terms, 28 million years ago Earth was in the Oligocene epoch, the Alps and the Himalayas were rising from the collision of India and Eurasia and large browsing mammals first appeared, along with the first monkey-like primates. During the succeeding 28 million years, whales, apes, large carnivores, and hominoids all appeared, while genetic drift was poking along removing one-half of the genetic variation.

Another property of genetic drift that is easy to derive is the probability that the A_1 allele will be the sole surviving allele in the population. This probability is called the *fixation probability*. The A_1 allele was fixed in two of the four replicate populations in which a fixation occurred. We could use the simulation to guess that the fixation probability of the A_1 allele is about one-half. In fact, the fixation probability is 0.2, as will emerge from a few simple observations.

As all variation is ultimately lost, we know that eventually one allele will be the ancestor of all of the alleles in the population. As there are $2N$ alleles in the population, the chance that any particular one of them is the ancestor of all (once $H = 0$) is just $1/(2N)$. If there were, say, i copies of the A_1 allele, then the chance that one of the i copies is the ancestor is $i/(2N)$. Equivalently, if the frequency of the A_1 allele is p, then the probability that all alleles are ultimately A_1 is p. In this case, we say that the A_1 allele is fixed in the population. Thus, the probability of ultimate fixation of a neutral allele is its current frequency, $\pi(p) = p$, to introduce a notation. This is as trivial as it is because all alleles are equivalent; there is no natural selection.

Notice that our study of H and the fixation probability agrees very well with the observations made earlier on the population composed of a single individual. You should use the simple case whenever your intuition for the more complicated case needs help.

Genetic drift is an evolutionary force that changes both allele and genotype frequencies. No population can escape its influence. Yet it is a very weak evolutionary force in large populations, prompting a great deal of debate over its relative importance in evolution. Drift is undeniably important for the dynamics of rare alleles, in small subdivided populations with very low migration rates, and in the neutral theory of molecular evolution. Beyond these three arenas, there is little agreement about the importance of drift.

Genetic drift appears to call into question the validity, or at least the utility, of the Hardy-Weinberg law. However, this is not the case except in the smallest of populations. The attainment of Hardy-Weinberg frequencies takes only a generation or two. Viewed as an evolutionary force, random mating has a time scale of one or two generations. Drift has a time scale of $2N$ generations, vastly larger than one or two for natural populations. When two forces have such different time scales, they rarely interact in an interesting way. This is certainly true for the interaction of drift and random mating. In any particular generation, the population will appear to be in Hardy-Weinberg equilibrium. The deviation of the frequency of a genotype from the Hardy-Weinberg expectation will be no more than about 1/(2N), certainly not a measurable deviation. Moreover, the allele frequency will not change by a measurable amount in a single generation. Thus, there is nothing that an experimenter could do to tell, based on allele and genotype frequencies, that the population does not adhere faithfully to the Hardy-Weinberg predictions.

Mutation and Drift

If genetic drift removes variation from natural populations, why aren't all populations devoid of genetic variation? The answer, of course, is that mutation restores the genetic variation that genetic drift eliminates. The interaction between drift and mutation is particularly important for molecular population genetics and its neutral theory. The neutral theory claims that most of the DNA sequence differences between alleles within a population or between species are due to neutral mutations. The mathematical aspects of the theory will be developed in this section. The following section will bring the mathematics and the data together. Mutation introduces variation into the population at a rate $2Nu$, where u is the mutation rate to neutral alleles. Genetic drift gets rid of variation at a rate $1/(2N)$. At equilibrium, the probability that two alleles different by origin are identical by state is given by the classic formula

$$\hat{G} = \frac{1}{1+4Nu} \qquad \text{...(6)}$$

There are many ways to obtain Formula 6. We will use a traditional approach that follows directly from Equation 1 with the addition of mutation.

The probability that a mutation appears in a gamete at the locus under study is u, which is called the mutation rate even though mutation probability would be a more accurate term. When a mutation does occur, it is assumed to be to a unique allele, one that differs by state from all alleles that have ever existed in the population. A population that has been around for a long time will have seen a very large number of different alleles. Consequently, our model of mutation is often called the infinite-allele model. The infinite-allele model is meant to approximate the large, though finite, number of alleles that are possible at the molecular level.

The value of G after one round of random mating and mutation, as a function of its current value, is

$$G' = (1-u)^2\left[\frac{1}{2N} + \left(1-\frac{1}{2N}\right)G\right]. \qquad \text{...(7)}$$

Notice that this formula differs from Equation 1 only by the factor $(1 - u)^2$ on the right side. This factor is the probability that a mutation did not occur in either of the two alleles chosen at random from the population in the "prime" generation. (The probability that a mutation did not occur in one allele is $1 - u$; the two alleles are independent.)

Equation 7 may be manipulated to obtain an approximation for ΔH. The approximation is based on the fact that mutation rates are small (often 10^{-5} to 10^{-10} depending on the context) and population sizes are large (often much larger than 10^4). The approximation is obtained in three steps. For the first two, approximate $(1 - u)^2$ with $1 - 2u$ and ignore terms with u/N as factors,

$$G' \approx (1-2u)\left[\frac{1}{2N} + \left(1-\frac{1}{2N}\right)G\right]$$

$$\approx \frac{1}{2N} + \left(1-\frac{1}{2N}\right)G - 2uG.$$

For the third step, set $H = 1 - G$ and rearrange a bit to get

$$H' \approx \left(1-\frac{1}{2N}\right)H + 2u(1-H).$$

The change in H in a single generation is

$$\Delta H \approx -\frac{1}{2N}H + 2u(1-H). \qquad \text{...(8)}$$

At equilibrium, $\Delta H = 0$. The value of H that satisfies $\Delta H = 0$ is

$$\hat{H} = \frac{4Nu}{1+4Nu}, \quad \text{...(9)}$$

which gives Equation 6 immediately.

Our route to the equilibrium value of H was rapid, as befits such a simple result. Now, we must return to the derivation to bring out a few features that will increase our understanding of the interaction of evolution forces. Examination of Equation 8 shows that the change in H is a sum of two components, which may be written as

$$\Delta H \approx \Delta_N H + \Delta uH.$$

The left component,

$$\Delta_N H = -\frac{1}{2N}H,$$

is the already familiar change in H due to genetic drift. The right component,

$$\Delta_u H = 2u\,(1 - H), \quad \text{...(10)}$$

is the change in H due to mutation and may be derived by considering the effects of mutation in isolation, that is, in an infinite population.

In an infinite population with mutation, the probability that two randomly chosen alleles are different in state in the next generation is

$$H' = H + (1 - H)\,[1 - (1 - u)^2];$$

which is the sum of the two ways that the two alleles may have come to be different. The first way is if their parent alleles were different by state, which occurs with probability H. The second way is if their parent alleles were not different and a mutation occurred in the production of at least one of the two alleles. The mutation term on the far right is one minus the probability that no mutation occurred, which is the same as the probability that at least one mutation occurred. By approximating $(1 - u)^2$ with $1 - 2u$ and performing some minor rearrangements, we recover Equation 10 for $\Delta_u H$.

The change due to drift is always negative (drift decreases genetic variation), whereas the change due to mutation is always positive (mutation increases genetic variation). Equilibrium is reached when these two evolutionary forces exactly balance, $-\Delta_N H = \Delta_u H$. The equilibrium is interesting only when $4Nu$ is moderate in magnitude. If $4Nu$ is very small, then genetic drift dominates mutation and genetic variation is eliminated from the population. If $4Nu$ is very large, mutation dominates drift and all $2N$ alleles in the population are unique in state.

The comparison of the strengths of evolutionary forces is key to an understanding of their interaction. The strength of a force may be quantified by the time required for the force to have a significant effect on the genetic structure of the population. In the previous section, we saw that the time required for genetic drift to reduce the heterozygosity of the population by one-half is proportional to the population size, N. We could say that the time scale of genetic drift is proportional to the population size. Similarly, the time scale of mutation is proportional to $1/u$. If $1/u \ll N$, the time scale of mutation is much less than that of drift, leading to a population with many unique alleles. If $N \ll 1/u$, the time scale of drift is shorter, leading to a population devoid of variation.

In addition to finding the equilibrium properties of the population, we can also derive the rate of substitution of neutral mutations. The rate of substitution could just as well have been called the rate of fixation, as it is a measure of how frequently genetic drift and mutation cause alleles to become fixed in the population. To calculate the rate of substitution, we need only multiply the average number of mutations that enter the population each generation by the fraction of those mutations that fix.

The average number of new mutations entering the population each generation is equal to the number of gametes produced each generation times the probability of a mutation in any one of them, $2Nu$. Thus, in each generation there will be, on average, $2Nu$ new mutations in the population. The simplicity of this result often obscures an important consequence: more mutations enter large populations each generation than enter small populations. One might expect evolution to proceed more rapidly in large than small populations. This is often the case, but not when dealing with neutral evolution, as we are in this chapter.

Of the new mutations that enter the population each generation, a fraction, $1/(2N)$, will fix, on average. This follows from our previous observation that the chance that any particular allele will fix in a population is equal to its frequency, which is $1/(2N)$ for a new mutation. Thus, the average rate of substitution, k, is $2Nu \times 1/(2N)$ or

$$k = u \qquad \text{...(11)}$$

In words, the rate of substitution of neutral alleles, k, is equal to the mutation rate to neutral alleles, u. This is one of the most remarkable results in all of population genetics. At first it offends our intuition. The fixation of alleles is caused by genetic drift. The strength of genetic drift depends on the population size. Our intuition wants the rate of fixation to depend on population size as well. However, the number of mutations entering the population each generation also depends on population size and does so in such a way as to cancel out drift's dependency on population size when deriving the rate of substitution.

There are a few subtleties in the interpretation of the neutral mutation rate, $k = u$. In a population that is so large that $4Nu \gg 0$, a new mutant allele will almost never fix in the population simply because a large number of new mutations enters the population each generation. If there are no fixations, how can Equation 11 be interpreted as the rate of substitution? To understand this paradox, we need an explicit model of a locus. One model in common use in population genetics, the infinite-sites, free-recombination model, imagines that a locus is a very large (effectively infinite) sequence of nucleotides, each evolving independently. Independent evolution requires that there is a fair amount of recombination between nucleotides, hence the "*free recombination*" part of the name. The most important consequence of having a large number of nucleotides is that the mutation rate per nucleotide is very small. For example, if the mutation rate at a locus of 1000 nucleotides were 10^{-6}, the mutation rate at each nucleotide would be 10^{-9}. With such a small mutational input per site, it is plausible that mutations at sites will fix, even if there is never a fixation of an allele at the locus. If we call the mutation rate per site u_s, then the rate of substitution per site is $k_s = u_s$. If there are n sites in the locus, then the rate of substitution for the locus is

$$k = nk_s = nu_s = u,$$

where u is the locus mutation rate.

The output of a computer simulation of the infinite-sites, free-recombination model shown. The vertical *axis* is the frequency of mutations at sites. During the course of the simulation, there were fixations at four sites. However, at no time was the population homozygous, so there could not have been any fixations of alleles.

The infinite-sites model may be immediately applied to DNA sequence data. For example, in the previous chapter we saw that *Drosophila melanogaster* and *D. erects* differed by 36 of 768 nucleotides. If we assume that the two species separated 23 million years ago and that there were exactly 36 substitutions on the lineages leading to the two species, then the rate of substitution per site is

$$k_s = \frac{36/768 \text{ substitutions/site}}{2 \times 23{,}000{,}000 \text{ years}}$$

$$\approx 10^{-9} \text{ substitution/site/year}$$

(The factor of two in the denominator is because the total time separating the two species is the sum of the lengths of the branching leading from their common ancestor to each species.) If we assume that the substitutions are neutral, then the neutral mutation rate must be $u_s = 10^{-9}$ mutations per site per year. If *Drosophila* has one generation per year, then the mutation rate per generation is also 10^{-9}. If, on the other hand, *Drosophila* has two generations per year, then the mutation rate per generation must be $(1/2) \times 10^{-9}$ so that the rate of neutral substitution matches the rate of substitution per year estimated from the sequence data.

Because the neutral rate of substitution depends only on the mutation rate, it is natural to expect that the rate of substitution will be approximately the same for different groups of species. In fact, the apparent constancy of evolutionary rates of proteins has long been used to argue that most evolution at the molecular level is due to drift and mutation alone. The neutral theory is concerned with this hypothesis.

Neutral Theory

The neutral theory of molecular evolution claims that most allelic variation and substitutions in proteins and DNA are neutral. Neutral evolution has been called non-Darwinian evolution because most substitutions are due to genetic drift rather than natural selection. However, the theory is not in conflict with Darwin's theory; rather, it simply claims that most substitutions have no influence on the survival of genotypes. Those few that do are subject to natural selection and change the adaptation of the species to its environment. The neutral theory was controversial when first proposed and remains controversial today. Many individuals proposed the theory, more or less independently, in the 1960s. Among these were N. Sueoka, E. Freese, A. Robertson, M. Kimura, T. Jukes, and J. L. King. The idea is so simple that many others undoubtedly thought of it at about the same time. The first paper to develop fully the population genetics aspects of the theory was the 1971 paper by Motoo Kimura and Tomoko Ohta, "Protein polymorphism as a phase of molecular evolution." This paper is a true classic in population genetics and will be examined in detail in this section.

The first section of the paper, "*Rate of Evolution*," points out that the rate of amino acid substitution per year is remarkably constant among vertebrate lineages for each protein examined. This constancy led to the concept of a molecular clock which, like the ticking of a

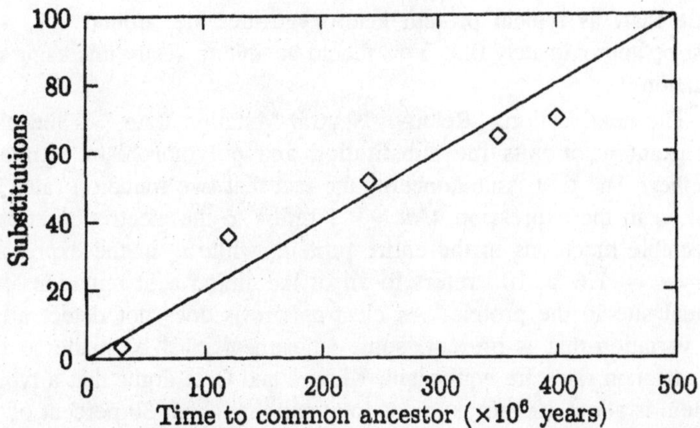

Fig. 9.4. The number of amino acid substitutions in beta globin that occurred in the lineages leading to humans and various species as a function of the time back to their common ancestors.

very slow clock, is the roughly constant rate of occurrence of substitutions through geologic time. Hemoglobins, experience about one amino acid substitution every one billion years at each amino acid site in vertebrate lineages. Thus, the rate of substitution in hemoglobins is approximately $k_s = 10^{-9}$ amino acid substitutions per amino acid site per year. Cytochrome *c* evolves more slowly, about 0.3 substitution per site per billion years, or $k_s = 0.3 \times 10^{-9}$. The average rate among the proteins examined was $k_s = 1.6 \times 10^{-9}$. The apparent constancy of the rate of substitution is viewed by Kimura and Ohta as strong evidence in favour of the neutral theory. They argue that natural selection, being more opportunistic and more subject to the vagaries of the environment, might be expected to cause a much more erratic pattern of substitutions.

If substitutions are neatral, we know from Equation 11 that the average neutral mutation rate, u_s, must be 1.6×10^{-9} amino acid mutations per amino acid site per year, which is strikingly close to nucleotide mutation rates as measured in the laboratory. Is this coincidence, or is amino acid evolution neutral with $k_s = u_s \approx 10^{-9}$? Kimura and Ohta show how polymorphism data may be used to support the hypothesis that most of the protein evolution is, in fact, neutral.

The next two sections of the paper, "Polymorphism in Subpopulations" and "Mutation and Mobility" deal with technical points and may be skipped on first reading. In the following section, "Heterozygosity and Probability of Polymorphism," Kimura and Ohta argue that, as typical protein heterozygosities are around 0.1, $4Nu$ must be approximately 0.1. You should be able to verify this using our Equation 9.

The next section, "Relative Neutral Mutation Rate," is the most important as it puts the substitution and polymorphism estimates together. The first issue concerns the fact that two mutation rates are used; u in the expression $4Nu \approx 0.1$ refers to the electrophoretically detectable mutations in the entire protein, while u_s in the expression $k_s = u_s = 1.6 \times 10^{-9}$ refers to all of the amino acid mutations at a typical site in the protein. As electrophoresis does not detect all of the variation that is present, some adjustment must be made so that the mutation rates are equivalent. Kimura and Ohta argue that a typical protein is about 300 amino acids long and that about 30 percent of the variation is detected by electrophoresis. Thus, the mutation rate to electrophoretically detectable variation for the entire protein is

$$u = 1.6 \times 10^{-9} \times 300 \times 0.3 = 1.44 \times 10^{-7}.$$

(Actually, their paper has an unfortunate typographical error that reports that $u = 1.6 \times 10^{-7}$.) The correct result, as given here, is viewed as an overestimate, so Kimura and Ohta use $u = 10^{-7}$ for the remainder of their investigation.

The electrophoretic mutation rate is in units of detectable mutations per protein per year, whereas Equation 9 for the equilibrium value of H requires the mutation rate per generation. The conversion is simple. For example, if mice have two generations per year, then the mutation rate per generation will be half of 10^{-7}. For mice, we have

$$4Nu = 4N \times 10^{-7}/2 \approx 0.1,$$

which gives $N \approx 5 \times 10^5$. The population size being estimated is called the *effective population size*. There was nothing in 1971 to suggest that the effective size of the house mouse was incompatible with about 10^5 individuals, despite our subjective impression that house mice exist in vast numbers throughout the world. Thus, the data on protein polymorphism and substitutions were viewed to be mutually compatible and to support the neutral theory.

Much has been written on the neutral theory since 1971. The theory slowly gained momentum through the 1970s, culminating in Kimura's book published in 1983. At that time, the neutral theory was the dominant explanation for most of protein and DNA evolution. More recently, the theory has fallen on hard times, particularly with regard to protein evolution. There are two aspects of the Kimura and Ohta paper that presented problems for the theory in later years. The first was the observation that the rate of substitution of amino acids is roughly constant per year. Our derivation of $k = u$ was done in time units of generations, not years. Thus, the rate of substitution should be roughly constant per generation rather than per year. As a consequence, creatures with shorter generation times should evolve faster than those with longer generation times. This generation-time effect, a clear prediction of the neutral theory, was not observed in proteins but was seen in noncoding DNA. This point, glossed over in the Kimura and Ohta paper, later caused a major shift in the neutral theory. The new version, due mainly to Tomoko Ohta, assumes that most amino acid substitutions are not neutral but are slightly deleterious.

The second point concerns the estimates of the effective population sizes using heterozygosity estimates. When the paper was written, most species that had been examined, from *Homo* to *Drosophila*, had heterozygosities in the narrow range of $0.05 < H < 0.15$. By implication, the effective population sizes of these species are also in a narrow range, in contradiction to the common belief that the population sizes of different species vary over several orders of magnitude. Compounding this problem is the fact that the effective sizes of some species, like *Drosophila*, might well be greater than 10^{10}. If the neutral theory were true, then the heterozygosity of *Drosophila* should be much higher than it is.

There are two refinements of the neutral theory that can explain the narrow range of heterozygosities. The first is Ohta's hypothesis that amino acid mutations are slightly deleterious and thus are less frequent than predicted under the strictly neutral model.

Effective Population Size

The model captured in Equation 1 contains explicit and implicit assumptions about the population that appear to compromise its usefulness. For example, the species is assumed to be hermaphroditic and randomly mating. A more egregious assumption is the constancy of the population size. Not only do real populations fluctuate in size, they often fluctuate wildly. Do these simplifying assumptions render our investigations irrelevant, or can we make some minor adjustments and continue with our hard-won insights about genetic drift? In many cases, we can make some adjustments by using the concept of the effective population size.

In most models of natural populations, no matter how complex, the heterozygosity eventually decreases geometrically, just as it does in our idealized population. However, the rate of decrease will no

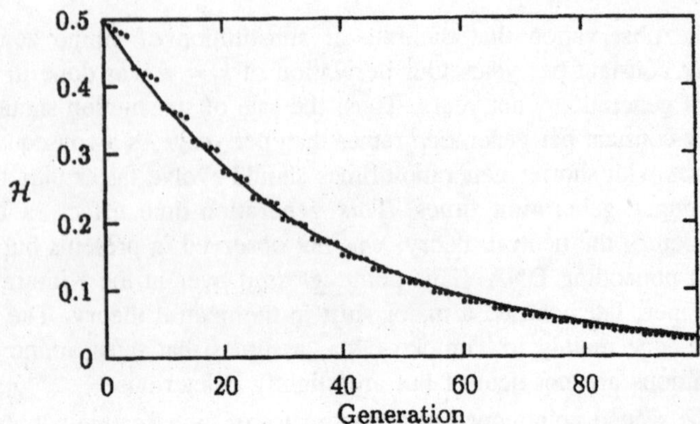

Fig. 9.5. The time course the H in two populations. The dots are from a population with changing size with a harmonic mean of 17.875.

longer be $1/(2N)$, but will be some new rate, call it $1/(2N_e)$, that depends on the particulars of the model. The parameter N_e is called the *effective size* of the population. It is the size of an idealized population whose rate of decay of heterozygosity is the same as that of the complicated population. Thus, we need only investigate how each complicating assumption influences the effective size of the population. From then on, we can simply substitute N_e for N in all of the preceding equations.

Of all the factors that affect the effective size, none is more important than fluctuations in the actual population size. Suppose, for example, the population sizes form a sequence N_1, N_2, ... indexed by the generation number. The value of H in generation $t + 1$, as a function of the population size in the previous generation, is

$$H_{t+1} = H_t\left(1 - \frac{1}{2N_t}\right).$$

Using the same argument that resulted in Equation 3, we obtain

$$H_t = H_0\prod_{i=0}^{t-1}\left(1 - \frac{1}{2N_i}\right).$$

At this point, we can use the definition of the effective population size to argue that the effective size of a population with fluctuating actual population sizes satisfies the equation

$$H_0\left(1 - \frac{1}{2N_e}\right)^t = H_0\prod_{i=0}^{t-1}\left(1 - \frac{1}{2N_i}\right). \qquad ...(12)$$

On the left side is H of an idealized population of actual size N_e and on the right side is H for the population complicated by population size fluctuations. The effective size of the complicated population is found by solving Equation 12 for N_e.

Equation 12 may be solved by using the techniques for approximating the product of terms that are close to one. The left side of Equation 12 is approximately $\exp(-t/2N_e)$, which is derived with the help of Equation 3. The right-hand side is covered by Equation 4. Together we get

$$e^{-t/2N_e} \approx e^{-\sum_{i=0}^{t-1} 1/2N_i}.$$

Equating the exponents and solving for the effective size, we have,

$$N_e \approx \left(\frac{1}{t} \sum_{i=0}^{t-1} \frac{1}{N_i} \right)^{-1}, \qquad \text{...(13)}$$

which is our final expression for the effective size of the population. You might recognize the formula for N_e as being the harmonic mean population size. The harmonic mean of the numbers $x_1, x_2, \ldots, x_n$ is the reciprocal of the average of the reciprocals of the x_i,

$$\frac{1}{\frac{1}{n}\left(\frac{1}{x_1} + \frac{1}{x_2} + \cdots + \frac{1}{x_n}\right)}.$$

One of the classic inequalities of mathematics states that the harmonic mean of a sequence of numbers is always less than or equal to the arithmetic mean.

The decay of H in a population whose size oscillates such that it is at 50 individuals for four generations, then 5 individuals for one generation, then 50 for four, and so forth. The figure also shows the decay of variation in a population of constant size 17.875, which is the harmonic mean population size of the oscillating population. The two populations clearly lose their variation at the same rate, even though they do not share exactly the same values for H.

The result that the effective size equals the harmonic mean population size is quite important because the harmonic mean is much more sensitive to small values than is the arithmetic mean. For example, suppose the population size is 1000 for nine generations and 10 for one generation. The arithmetic mean in this case is

$$\frac{9}{10} \times 1000 + \frac{1}{10} \times 10 = 901.$$

The harmonic mean is

$$\left(\frac{9}{10}\times\frac{1}{1000}\times\frac{1}{10}\times\frac{1}{10}\right)^{-1}=91.4$$

which is about an order of magnitude smaller than the arithmetic mean population size. Were species to experience occasional but recurring crashes in population size, called bottlenecks by population geneticists, then the effective sizes of species would be considerably smaller than their "typical" sizes.

Population size fluctuations can account for two of the problems with the neutral theory mentioned in the previous section. The first is the problem of too little molecular variation based on our sense of current population sizes. If the sizes of most species do fluctuate wildly, then their effective sizes may be close to 10^5 as required by the neutral estimation, even though their current sizes are several orders of magnitude larger. The second problem concerns the narrow range of heterozygosities. Recall that the harmonic mean is determined more by the population size during bottlenecks than during times of abundance. However, there is a limit to how small population sizes can be, as rare species inevitably face extinction. In fact, extant species are precisely those species that have not suffered recurring catastrophic crashes in population sizes. The net effect is to make the harmonic means of species more similar than their actual numbers at any point in time. Although we should still be uncomfortable with the narrow range of heterozygosities, we can be less uncomfortable given insights provided by our analysis of the effective size of species.

Other factors also affect the effective size, but not so dramatically as bottlenecks. In randomly mating dioecious species with different numbers of males, N_m, and females, N_f, the effective size is

$$N_e=\frac{4N_mN_f}{N_f+N_m}.$$

For example, suppose the number of males is a times the number of females, $N_m = \alpha N_f$. The ratio of the effective size to the actual size is then

$$\frac{N_e}{N_m+N_f}=\frac{4\alpha}{(1+\alpha)^2}.$$

In a species with one-tenth the number of males as females, the effective size will be about 33 percent of the actual population size. This is not a particularly extreme reduction. Except for species with population sizes so small that extinction seems imminent anyway, a 33 percent reduction in the effective size is of little consequence. Our estimates of actual population sizes are so imprecise that the adjustment required for such factors as different numbers of the sexes often seem unimportant compared to the very large reductions that come with fluctuations in population size.

COALESCENT

The parameter $\theta = 4Nu$ determines the level of variation under the neutral model. Assuming that the variation at a locus is neutral, θ may be estimated by using the observed homozygosity and Equation 6, as done, for example, by Kimura and Ohta. If the data come as DNA sequences rather than from protein electrophoresis, then it is possible to use the theory of coalescents and the additional information in the sequence to obtain a better estimate of θ.

A coalescent is the lineage of alleles in a sample traced backward in time to their common ancestor allele. Each of the alleles in the sample is descended from an allele in the previous generation. In a large population, the immediate ancestors of the sampled alleles are likely to be distinct. However, if the ancestors of the sampled alleles are followed backward for a long time, eventually a generation will be found with a common ancestor of two of the alleles. In the figure, the two central alleles have a common ancestor t_1 generations in the past. At t_1 we say that a coalescence occurred because the number of lineages in the coalescent is reduced by one. The next coalescence occurs t_2 generations in the past, and the final coalescence occurs t_3 generations in the past. Time, when discussing coalescents, is always measured in units of generations in the past.

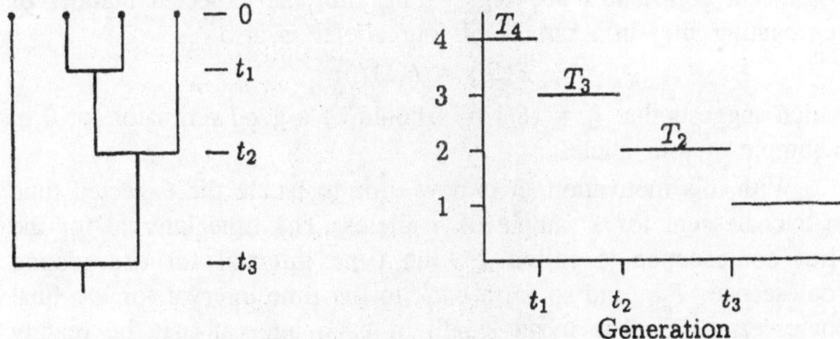

Fig. 9.6. On the left is an example of a coalescent for four alleles. On the right is a graph showing the size of the coalescent as a function of time measured backward.

The coalescent is sometimes called the genealogy of the sample because it captures the genealogic relationships of the sampled alleles. The coalescent cannot be known, of course, because there is no way to know which alleles share which common ancestors, nor is there any way to know the times of coalescence events. Coalescents are useful constructs because in natural populations mutations occur on the lineages and these mutations can be used to infer some of the coalescent and to estimate θ. Each mutation on the coalescent results in a segregating site in the sample. The number of segregating sites in a sample of n alleles, S_n, turns out to contain enough information to estimate θ.

Although we cannot know t_1, t_2, and t_3 for a sample of four alleles, it is possible to state that the total time in the coalescent, T_c, depends on these times through

$$\begin{aligned} T_c &= 4 \times t_1 + 3 \times (t_2 - t_1) + 2 \times (t_3 - t_2) \\ &= 4T_4 + 3T_3 + 2T_2, \end{aligned}$$

where T_i is the time required to reduce a coalescent with i alleles to one with i - 1 alleles.

For a neutral mutation rate of u, the expected number of mutations in the coalescent is $T_c u$. As we will soon see, the expected value of T_c, for a sample of four alleles is

$$E\{T_C\} = 4N\left(1+\frac{1}{2}+\frac{1}{3}\right) = \frac{44N}{6},$$

which implies that, on average, there will be

$$uE\{T_c\} = \theta(11/6)$$

mutations in a sample of four alleles. Because each mutation in the coalescent contributes one segregating site, the expected number of segregating sites in a sample of four alleles is also

$$E\{S_4\} = \theta(11/6),$$

which suggests that $\hat{\theta} = (6/11)S_4$ should be a good estimator for θ in a sample of four alleles.

With this motivation, it is now time to tackle the expected time in a coalescent for a sample of n alleles. The time interval for the first coalescence is called T_n, the time interval for the second coalescence, T_{n-1}, and so forth back to the time interval for the final coalescence, T_2. The mean length of each interval may be readily found if we know the probability that a coalescence does not occur in the previous generation or, as we prefer to call it, the next ancestral generation.

Consider the history of the n alleles in turn. The first allele has, of course, an ancestor allele in the first ancestral generation. The second allele will have a different ancestor allele with probability

$$1-\frac{1}{2N}=\frac{2N-1}{2N}.$$

The right-hand side of this equation is particularly informative because it shows that there are a total of $2N$ possible ancestors for the second allele, but only $2N - 1$ that are different from the first allele. The probability that the third allele does not share an ancestor with the first two, assuming that the first two do not share an ancestor, is $(2N - 2)/(2N)$; the total probability that the first three do not share an ancestor is

$$\frac{2N-1}{2N}\times\frac{2N-1}{2N}.$$

With an obvious leap of intuition, the probability that the n alleles all have different ancestors is

$$\left(1-\frac{1}{2N}\right)\left(1-\frac{2}{2N}\right)\cdots\left(1-\frac{n-1}{2N}\right)\approx 1-\frac{1}{2N}-\frac{2}{2N}-\cdots-\frac{n-1}{2N},$$

where terms of order N^{-2} and smaller have been ignored in the approximation. The probability that a coalescence occurs is one minus the probability that it does not, or

$$\frac{1+2+\cdots+(n-1)}{2N}=\frac{n(n-1)}{4N},$$

where the final step uses the fact that the sum of the first m integers is

$$1+2+\cdots+m=m(m+1)/2.$$

If the probability that the first coalescence occurs in any particular generation is $n(n - 1)/(4N)$, then the probability distribution of the time until the first coalescence is a geometric distribution with probability of success $p = n(n - 1)/(4N)$. The mean of the geometric distribution is the reciprocal of the probability of success, giving

$$E\{T_n\}=\frac{4N}{n(n-1)}.$$

The beauty of the coalescent approach is that all of the hard work is now over. A little reflection shows that the mean time interval leading from a coalescent with i alleles to one with $i - 1$ alleles is just

$$E\{T_i\}=\frac{4N}{i(i-1)}, \qquad \text{...(14)}$$

as there is nothing special about n in the previous derivation of ET_n.

The total time in all of the branches of a coalescent is

$$T_c = \sum_{i=2}^{n} iT_i,$$

which, using the fact that the expectation of the sum of random quantities is the sum of the expectations of those quantities, is

$$E\{T_c\} = \sum_{i=2}^{n} iE\{T_i\} = 4N\sum_{i=2}^{n}\frac{1}{i-1}.$$

Recalling that the expected number of segregating sites is the neutral mutation rate, u, times the expected time in the coalescent, we have

$$E\{S_n\} = uE\{T_c\} = \theta\sum_{i=2}^{n}\frac{1}{(i-1)},$$

which suggests that

$$\hat{\theta} = \frac{S_n}{1+\frac{1}{2}+\frac{1}{3}\cdots+\frac{1}{n-1}} \qquad \text{...(15)}$$

should be a good estimator for $\theta = 4Nu$.

For example, there were $S_{11} = 14$ segregating sites in Kreitman's *ADH* sample. The denominator of the right-hand side of Equation 15 for $n = 11$ is 2.93, so the estimate of $4Nu$ is $\theta = 4.78$. Theta for a nucleotide site, rather than for the entire locus, is 4.78/768 = 0.0062. These estimates must be viewed with some skepticism because they require that the neutral model accurately reflect the evolutionary dynamics of the *ADH* locus and that the population be in equilibrium. Most population geneticists have reservations about the neutrality of replacement mutations, but many do accept the neutrality of silent mutations. Thus, you will frequently see θ estimated for silent variation only. In the case of *ADH,* this involves including only the 13 segregating sites with silent variation.

The coalescent may be used to derive the probability that two alleles different by origin are different by state, $\hat{H}$. The two alleles will be different if a mutation occurred on the lineages leading from their common ancestor; otherwise they will be identical. As the lineages of the two alleles are traced backward in time, either a coalescence or a mutation will occur first. The probability that a coalescence occurs in any particular generation is $1/(2N)$, and the probability that a mutation occurs is

$$1 - (1 - u)^2 \approx 2u.$$

The probability that a mutation is the first event to occur is just its relative probability of occurrence,

$$\frac{2u}{2u+1/(2N)} = \frac{4Nu}{1+4Nu},$$

which is the same as Equation 9. This derivation of $\hat{H}$ using a coalescent argument is both easier and more instructive than the difference equation approach used previously and hints at the power of the coalescent for solving ostensibly difficult problems. The coalescent has become an indispensable construct for the analysis sequence data in population genetics. Dick Hudson, one of the pioneers in the application of coalescent theory, has written an excellent paper describing its use.

Binomial Sampling

The description of genetic drift was based on the probability of identity by state, *G*. Although *G* is perfect for describing the average rate of decay of variability, it does not give a good feeling for the underlying randomness of genetic drift. In fact, the equation

$$H_t = H_0\left(1-\frac{1}{2N}\right)^t$$

could give the impression that the heterozygosity of any particular population decreases nonrandomly. Nothing could be further from the truth, which uses the same data but plots heterozygosities rather than allele frequencies. Note that the heterozygosity of any particular population does not decrease monotonically. Rather, it jumps up and down, eventually hitting zero and staying there. After staring, for a while, it is reasonable to start wondering exactly what H_t really is. A promising conjecture is that H_t is the average heterozygosity of a very large number of replicate populations. Not a bad conjecture, but it is off by a factor of $1 - 1/(2N)$.

The short algorithm used for the computer simulations at the beginning of this chapter also happens to be an accurate description of the way most population geneticists model genetic drift. The heart of the algorithm is the copying of alleles from the current generation into the next generation, which is repeated $2N$ times. The probability that i A_1 alleles make it into the next generation is the binomial probability

$$\text{Prob}\{i\ A_1 \text{ allele}\} = \frac{(2N)}{i!(2N-i)!}p^i q^{2N-i},$$

where i can be 0, 1, … $2N$. The binomial distribution describes the probability of i successes in n independent experiments, where the probability of success in any one experiment is p. With genetic drift, the experiment is repeated $2N$ times and the probability of success—the probability of copying an A_1 allele—is the allele frequency p.

The mean of a binomial random variable with parameters n and p is np. Thus, the mean number of A_1 alleles to appear in the next generation is

$$E\{i\} = 2Np.$$

The allele frequency in the next generation is $i/2N$. The mean allele frequency is

$$E\{i/2N\} = E\{i\}/2N = p,$$

by Equation 9. In other words, the mean allele frequency does not change under genetic drift.

A more complete description of the mean change in p is

$$E\{\Delta_N p|p\} = E\{i/2N - p|p\} = E\{i/2N|p\} - p = 0.$$

This stream of equalities uses a a new symbol, $|$, which stands for given, making the expectation a conditional expectation. For example, $E\{\Delta p|p\}$ is the mean change in p given its current value, p. We need this device because we cannot know with certainty what p will be in any particular generation as it is changing at random each generation. But, if we were told its value in a particular generation, then we could say something about its value in the next generation. The $|$ symbol is shorthand for being told the current value of p.

The variance of the change in the allele frequency may be found by a similar argument,

$$\text{Var}\{\Delta p|p\} = \text{Var}\{i/2N - p|p\} = \frac{pq}{2N}.$$

The last step is interesting. First, note that the conditioning on p makes the expression $i/2N - p$ the difference between a random quantity, $i/2N$, and a nonrandom quantity, p. (p is not random because of the conditioning on its value.) We know that

$$\text{Var}\{i/2N - p\} = \text{Var}\{i\}/(2N)^2$$

when p is a constant. As $\text{Var}\{i\} = 2Npq$, we obtain our important result

$$\text{Var}\{\Delta p|p\} = \frac{pq}{2N}. \qquad \text{...(16)}$$

All of the above shows that the number of A_1 alleles in the daughter generation is binomially distributed and that the mean and variance of the change in the allele frequency are

$$E\{\Delta p\} = 0 \qquad \text{...(17)}$$

$$\mathrm{Var}\{\Delta p\} = \frac{pq}{2N}. \qquad ...(18)$$

The fact that the variance is inversely proportional to the population size means that the dispersion of the allele frequency around the current value will be less in larger populations. The actual distributions of allele frequencies in the daughter generation for population sizes of 10 and 100 are illustrated in figure. The narrow range of likely p values for the larger population size is apparent.

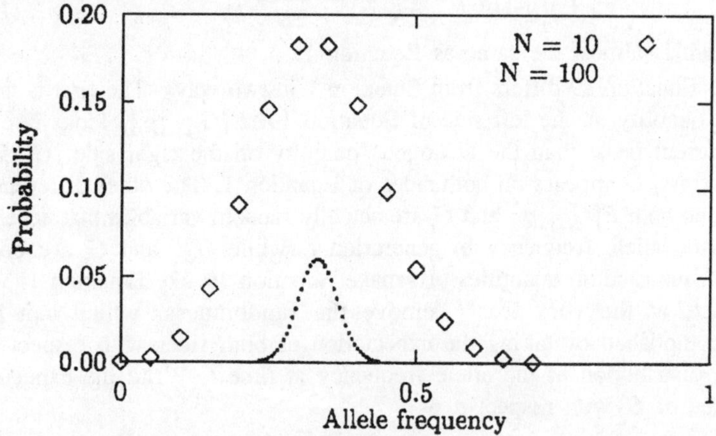

Fig. 9.7. The probabilities of allele frequencies after one round of random mating for population sizes of N = 10 and N = 100 and initial allele frequency p = 1/3.

How can we connect this development of genetic drift with the previous one based on G? Let's begin with some notation. Call the allele frequencies in the tth generation p_t and q_t and the homozygosity

$$G_t = p_t^2 + q_t^2.$$

Our conjecture that G_t is equal to the mean of G_t suggests that a natural place to begin our investigation is with the expected value of G_{t+1} given its value in generation t. Rather than conditioning on the current value of the homozygosity, it is somewhat easier to condition on the allele frequency in the current generation, which amounts to the same thing,

$$E\{G_{t+1}|pt\} = E\{p_{t+1}^2|pt\} + E\{q_{t+1}^2|qt\}$$

$$= \frac{p_t q_t}{2N} + p_t^2 + \frac{p_t q_t}{2N} + q_t^2$$

$$= \frac{1}{2N} + (1 - \frac{1}{2N})(p_t^2 + q_t^2)$$

$$= \frac{1}{2N} + (1 - \frac{1}{2N})G_t.$$

In verifying these steps, recall that

$$\mathrm{Var}\{p\} = E\{p^2\} - E\{p\}^2$$

and that

$$2pq = 1 - p^2 - q^2$$

At this junction, we have

$$E\{G_{t+1}|pt\} = \frac{1}{2N} + (1 - \frac{1}{2N})G_t, \quad \text{...(19)}$$

which is almost the same as Equation 1.

Equation 19 differs from Equation 1 in two ways. The first is that the quantity on the left side of Equation 19, $E\{G_{t+1}|p_t\}$, looks like a different beast than the analogous quantity on the right side, G_t. By contrast, G appears on both sides of Equation 1. The other difference is that both $E\{G_{t+1}|p_t\}$ and G_t are actually random variables that depend on the allele frequency in generation t, while G_{t+1} and G_t are both both nonrandom quantities. To make Equation 19 like Equation 1, we must, at the very least, remove the randomness, which can be accomplished by taking the expectation of both sides with respect to the distribution of the allele frequency at time t. Write the expected value of G_t with respect p_t as

$$E_{pt}\{G_t\} = \bar{G}_t$$

and that of G_{t+i} as

$$\begin{aligned} E_{pt}\{E\{G_{t+1}|p_t\}\} &= E\{E_{pt}\{G_{t+1}|p_t\}\} \\ &= E\{\bar{G}_{t+1}\} \\ &= \bar{G}_{t+1}. \end{aligned}$$

The first line uses the fact that the order in which expectations are taken does not matter. The second line uses the definition of $\bar{G}$. The third line is trivial as the expected value of a constant, $\bar{G}_{t+1}$ in this case, is just the constant. Substituting these results into Equation 19 gives

$$\bar{G}_{t+1} = \frac{1}{2N} + (1 - \frac{1}{2N})\bar{G}_t,$$

which is the same as Equation 1 and whose solution is

$$\bar{H}_t = 1 - \bar{G} = H_0\left(1 - \frac{1}{2N}\right)^t. \quad \text{...(20)}$$

We do not need a line over H_0 because the initial condition is a fixed quantity, so there is no need to take its mean.

Just because G_t and $\overline{G}_t$ satisfy the same equation does not mean that they are equal. The solution to a dynamical equation like Equation 1 depends on the initial value of the state variable. You should have solved problem by now and should know that

$$G_0 = \frac{1}{2N} + \left(1 - \frac{1}{2N}\right)G_0,$$

or, equivalently, that

$$H_0 = \left(1 - \frac{1}{2N}\right)H_0.$$

Plugging this into Equation 20 gives

$$\overline{H}_t = H_0\left(1 - \frac{1}{2N}\right)^{t+1},$$

which differs from the value of H_t given in Equation 3 by the factor 1 - 1/2*N*. For large populations, this factor will be so close to one that $G_t \approx \overline{G}_t$. Finally, we are able to conclude with confidence that the expected heterozygosity of a population decreases at the rate 1/2*N*.

While the amount of work needed to reach a rather simple conclusion might seem excessive, it is work well spent. Not only did we learn more about the stochastic nature of genetic drift, but we also put ourselves in a position to solve some more important problems, like the fixation probability of a selected mutation.

10

NATURAL SELECTION

The results of natural selection, the evolutionary force most responsible for adaptation to the environment, are evident everywhere, yet it is remarkably difficult to observe the time course of changes brought about by selection The reason, of course, is that most evolutionary change is extraordinarily slow. Significant changes in the frequencies of genotypes take longer than the lifetime of a human observer. This temporal imbalance is the greatest obstacle to the study of evolution and is the main reason why so much of our understanding of evolutionary processes comes from theoretical and mathematical arguments rather than direct observation, as is typical in other areas of biology.

Occasionally, we are able to observe natural selection in action either because the strength of selection is so great that change occurs very quickly or because the organism, perhaps a bacteria or virus, has a very short generation time. The European scarlet tiger moth, *Panaxia dominula*, provides one well-studied example. In a population just outside of Oxford, England, an allele that reduces the spotting on the forewing, the *medionigra* allele, is found in fairly high frequency. As this allele is found nowhere else, it has attracted attention from butterfly enthusiasts. The frequency of the *medionigra* allele declined fairly steadily from 1939 until 1955, after which it began hopping around erratically. Although the complete record is difficult to interpret, the period of steady decline appears to be a case of natural selection preferring the common allele over the *medionigra* allele. If so, how strong is the selection? Other questions come to mind as well. Why is the *medionigra* allele less fit? If it is less fit, how did it get to a

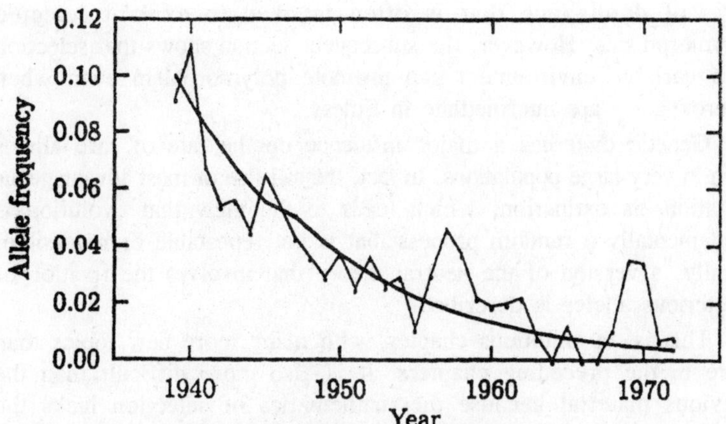

Fig. 10.1. The observed frequency of the medionigra allele in the scarlet tiger moth population compared to the expected frequency assuming a 10 percent disadvantage.

frequency of 10 percent before beginning its decline? While we will not be able to provide complete answers to any of these questions, we will be able to discuss them much more intelligently after a theoretical investigation of the nature and consequences of natural selection.

In this chapter, we will discover how natural selection changes allele frequencies by examining some one-locus models of selection. Natural selection works when genotypes have different fitnesses. To a geneticist, fitness is just another trait with a genetic component. To an evolutionist, it is the ultimate trait because it is the one upon which natural selection acts. Fitness is a complicated trait, even in the context of a simple one-locus, two-allele model. There is individual fitness, genotype fitness, relative fitness, and absolute fitness. We will spend some time making these different aspects of fitness clear before tackling the problem of the dynamics of natural selection.

An examination of the dynamics of natural selection quickly leads to the conclusion that the dominance relationships between alleles affecting fitness have a profound affect on the outcome of selection. Fortunately, the dominance of fitness alleles can be investigated experimentally; a study of viability in *Drosophila melanogaster* populations is described. A major conclusion of this study is that there is an inverse homozygous-heterozygous effect for deleterious alleles: alleles that have large deleterious effects when homozygous tend to be nearly recessive, whereas alleles with small homozygous effects tend to be nearly additive. A casualty of the study is overdominance, the form of dominance that is often invoked to explain selected polymorphisms. However, the subsequent section shows that selection in a variable environment can promote polymorphism even when heterozygotes are intermediate in fitness.

Genetic drift has a major influence on the fate of rare alleles even in very large populations. In fact, the sad fate of most advantageous mutations is extinction, which leads to the view that evolution is fundamentally a random process that is not repeatable or reversible. Finally, a version of the neutral theory that involves the fixation of deleterious alleles is described.

This is an ambitious chapter, with many more new topics than were in the preceding chapters. It is also more difficult than the previous material because the mathematics of selection lacks the elegance and simplicity of the mathematics of drift and mutation.

Fundamental Model

Natural selection is most easily studied in the context of an autosomal locus in a hermaphroditic species whose life cycle moves through a synchronous cycle of random mating, selection, random mating, selection, and so forth. Our entrance into the cycle is with newborns produced just after a round of random mating by their parents. Figure shows that the frequency of the A_1 allele among the newborns is called p, which is the same as the allele frequency in their parents. As their parents mated at random, the genotype frequencies of the newborns will conform to Hardy-Weinberg expectations.

The newborns must survive to adulthood in order to reproduce. The probability of survival of an individual will, in general, depend on the genotype of the individual. Let the probabilities of survival or, as they are more usually called, the viabilities of A_1A_1, A_1A_2, and A_2A_2 individuals be w_{11}, w_{12}, and w_{22}, respectively. Viabilities may be thought of as either probabilities of survival of individuals or the fraction of individuals that survive. The latter allows us to see immediately the consequences of selection because the frequency of a genotype after

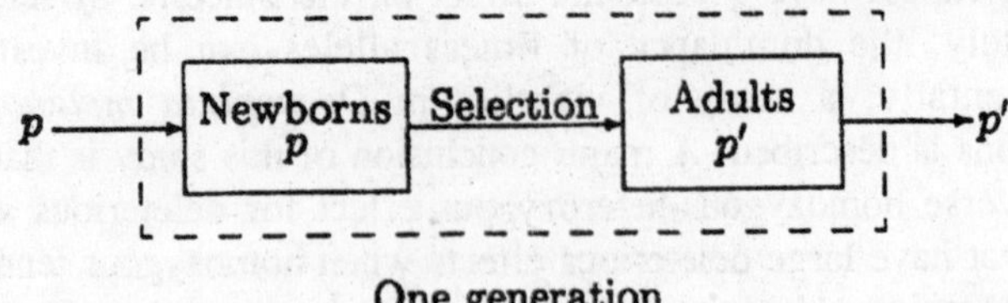

Fig. 10.2. The simple life cycle used in the fundamental model of selection.

selection is proportional to its frequency before selection times its viability, or

frequency after selection $\propto$ newborn-frequency $\times$ viability.

For example, the frequency of A_1A_1 in the adults is proportional to p^2w_{11}.

To obtain the relative frequencies of the three genotypes in the adults, we must find a constant of proportionality such that the sum of the three genotype frequencies in the adults is one. The following worksheet shows how this is done.

Genotype:	A_1A_1	A_1A_2	A_2A_2
Frequency in newborns:	p^2	$2pq$	q^2
Viability:	w_{11}	w_{12}	w_{22}
Frequency after selection:	$p^2w_{11}/\bar{w}$	$2pqw_{12}/\bar{w}$	$q^2w_{22}/\bar{w}$.

The constant of proportionality,

$$\bar{w} = p^2w_{11} + 2pqw_{12} + q^2w_{22},$$

is chosen such that

$$\frac{p^2w_{11}}{\bar{w}} + \frac{2pqw_{12}}{\bar{w}} + \frac{q^2w_{22}}{\bar{w}} = 1,$$

as required. The quantity $\bar{w}$ has special meaning in population genetics. It is called the mean fitness of the population.

After selection, the frequency of the A_1 allele may have changed. The new allele frequency, p', is

$$p' = \frac{p^2w_{11} + pqw_{12}}{\bar{w}}.$$

(Don't forget that each heterozygote has only one A_1 allele.) The change in the frequency of the A_1 allele in a single generation, $\Delta_s p = p' - p$, follows from

$$p' - p = \frac{p^2w_{11} + pqw_{12} - p\bar{w}}{\bar{w}}$$

$$= \frac{p[pqw_{11} + q(1-2)w_{12} - q^2w_{22}]}{\bar{w}},$$

which simplifies to

$$\Delta_s p = \frac{pq[p(w_{11} - w_{12}) + q(w_{12} - w_{22})]}{p^2w_{11} + 2pqw_{12} + q^2w_{22}} \quad \text{...(1)}$$

This is probably the single most important equation in all of population genetics and evolution! Admittedly, it isn't pretty, being a ratio of two polynomials with three parameters each. Yet, with a little poking around, this equation easily reveals a great deal of the dynamics of natural selection.

Relative Fitness

Notice that the terms in the numerator and denominator of Equation 1 all have a viability as a factor. Thus, if we were to divide the numerator and denominator by a viability, say by w_{11}, every viability in Equation 1 would become a ratio of that viability and w_{11}, yet the numerical value of Δp would not change at all. In other words, we could use as our definition of viability either the original definition based on absolute viabilities or a new one based on the relative viabilities of genotypes when compared to one particular genotype.

Genotype:	A_1A_1	A_1A_2	A_2A_2
Viability	w_{11}	w_{12}	w_{22}
Relative viability:	1	w_{12}/w_{11}	w_{22}/w_{11}

In either case, the dynamics of selection as captured in Equation 1 are the same. An important insight in its own right, this is also of great utility as it allows a much more informative choice of parameters than the w_{ij}'s used thus far.

Up to this point, w_{ij} has been called the viability of genotype A_iA_j. More often, w_{ij} is called the fitness, or sometimes the absolute or Darwinian fitness, of genotype A_iA_j. In nature, the fitness of a genotype has many components including viability, fertility, developmental time, mating success, and so forth. Most of these components, other than viability, cannot be included in a simple model like that defined by Equation 1. Yet, if the differences in fitnesses between genotypes are small, Equation 1 is a good approximation to the actual dynamics as long as the values of the w_{ij} are chosen appropriately. Here, we will not investigate these more complicated models but will from this point on refer to w_{ij} as a fitness and allow it to take on any values greater than or equal to zero. As the dynamics of selection depend on relative fitnesses, nothing really changes by allowing this broadened scope for w_{ij}.

A common notational convention for relative fitnesses is

Genotype:	A_1A_1	A_1A_2	A_2A_2
Relative fitness:	1	$1 - hs$	$1 - s$

where $1 - hs = w_{12}/w_{11}$ and $1 - s = w_{22}/w_{11}$.

The parameter s is called the selection coefficient. It is a measure of the fitness of A_2A_2 relative to that of A_1A_1. If the selection coefficient is positive, A_2A_2 is less fit than A_1A_1; if it is negative, A_2A_2 is more fit. In most of what follows, we will assume that the selection coefficient is in the range $0 \leq s \leq 1$. Nothing is lost in doing this because the labeling of alleles is completely arbitrary. The symbol A_1 will usually be attached to the allele whose homozygote is more fit than the other homozygote.

The parameter h is called the heterozygous effect. It is a measure of the fitness of the heterozygote relative to the selective difference between the two homozygotes. As such, it is really a measure of dominance, as shown in the following table.

$h = 0$	A_1 dominant, A_2 receissive
$h = 1$	A_2 dominant, A_1 recessive
$0 < h < 1$	incomplete dominance
$h < 0$	overdominance
$h > 1$	underdominance

Only the cases of incomplete dominance, overdominance, and underdominance are of general evolutionary interest. The cases of complete dominance given in the first two lines of the table are regarded as special cases that are unlikely to occur for any pair of naturally occurring alleles. Even classic cases of complete dominance like "recessive" lethals in human populations are now thought to be cases of incomplete dominance with $s = 1$ and h small, say about 0.01, but definitely greater than zero. The case $h = 1/2$ is of evolutionary importance as many alleles with a very small effect on fitness are close to additive; that is, close to the situation where the heterozygote is exactly intermediate between the two homozygotes.

Equation 1 for the single-generation change in the frequency of the A_1 allele, using relative frequencies, becomes

$$\Delta_s p = \frac{pqs[ph + q(1-h)]}{\bar{w}}. \qquad ...(2)$$

(To obtain this equation, simply make the substitutions $w_{11} = 1$, $w_{12} = 1 - hs$, and $w_{22} = 1 - s$ in equation 1.) The mean fitness becomes

$$\bar{w} = 1 - 2pqhs - q^2 s \qquad ...(3)$$

when using relative fitnesses. Not only is Equation 2 simpler than Equation 1, it is also more suggestive of the dynamics. We will soon see that h determines where the allele frequency ends up and s determines how quickly it gets there.

Thus far, we have discussed absolute and relative fitness, but not individual fitness, which is, as its name implies, a property of an

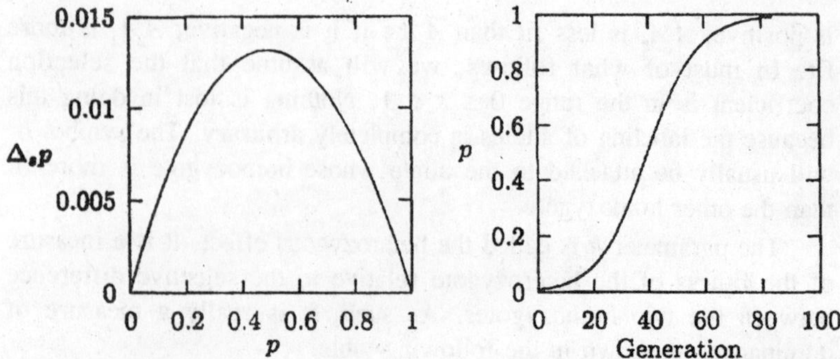

Fig. 10.3. Properties of directional selection with $h = 0.5$ and $s = 0.1$. The left-hand graph shows the change in the allele frequency in a single generation. The right-hand graph show the evolution of the allele frequency over 100 generations.

individual as opposed to a genotype. If the only factor determining fitness were viability, then an individual's fitness would be either zero or one, depending on whether or not the individual survived to reproduce. The absolute fitness of a particular genotype is simply the average of the individual fitnesses of all those individuals with that genotype. For example, if the probability that an A_2A_2 individual's fitness is one equals 0.9 and is zero equals 0.1, then the absolute fitness of the A_2A_2 genotype is

$$0.9 \times 1 + 0.1 \times 0 = 0.9.$$

Three Kinds of Selection

Equation 2 may be used to solve the following problem: Given the initial frequency of the A_1 allele and the parameters *s* and *h*, what will be the ultimate fate of the A_1 allele in the population? Will it take over in the population ($p \to 1$), disappear from the population ($p \to 0$), approach some intermediate value ($p \to \hat{p}$), or not change at all? As we shall see, all four outcomes are possible. Which one prevails depends on the dominance relationships between alleles and on the initial frequency of the allele.

The type of selection that Darwin had in mind in *On the Origin of Species* (1859) is called directional selection. Directional selection occurs with incomplete dominance ($0 < h < 1$). The fitness of A_1A_1 exceeds that of A_1A_2, which, in turn, exceeds that of A_2A_2. It should come as no surprise that *p* continually increases or, equivalently, that $\Delta_s p > 0$. The change in the allele frequency in a single generation, $\Delta_s p$, as a function of *p* is illustrated in the left-hand side. Because $s > 0$, Equation 2 shows that the sign of $\Delta_s p$ is determined by

$$ph + q(1 - h),$$

which is always positive with incomplete dominance ($0 < h < 1$). Thus, the allele frequency always increases no matter what its current value and, as a consequence, $p \rightarrow 1$, as illustrated in the right-hand. The rate of change of p is strongly dependent on p itself. Evolution by natural selection proceeds very slowly when there is little genetic variation; that is, when p is close to zero or one. Selection is most effective when genetic variation is near its maximum, $p = 1/2$.

Note that p will not equal one in a finite number of generations. As p gets closer to one, Δp approaches zero fast enough to prevent $p = 1$ in finite time. However, in finite populations genetic drift dominates selection when p gets very close to one and will thus cause fixations to occur in finite time.

The decrease in the frequency of the *medionigra* allele in *Panaxia* from 1939 until about 1955 may well be due to directional selection. If the selection coefficient for the *medionigra* allele were $s = 0.1$, the heterozygous effect $h = 1/2$, and the initial frequency $q = 0.1$, then the allele frequency trajectory obtained from Equation 2 fits the observed allele frequency quite well. Does this constitute a "proof" that direction selection is responsible for the allele frequency change? Unfortunately, no. Many different forms of selection have trajectories that fit the data as well as does directional selection. A convincing demonstration that directional selection is operating requires direct measurements of the selection coefficient to show that $0 < s < 1$ and of the heterozygous effect to show that $0 < h < 1$. The obstacles to obtaining these estimates in a natural setting are almost insurmountable. Not only are there the obvious problems of determining the viability, fertility, mating success, longevity, and so on of genotypes with sufficient accuracy, there is also the problem of demonstrating that the measured parameters are due to alleles at the *medionigra* locus and not to those at some closely linked locus. A good introduction to the problems of measuring selection in nature is John Endler's book, *Natural Selection in the Wild* (1986).

The entire history of the *medionigra* allele is a mystery. The allele is found only in the Oxford population, where it appears to have been at a selective disadvantage from at least 1939 to 1955. If it were always at a disadvantage, then why was its frequency as high as 10 percent in 1939? An obvious answer, one that fits with the standard Darwinian view of the world, is that the environment has changed. The fitness of a genotype is determined by its adaptation to the environment in which it finds itself. Even a casual observer notices that environments are constantly changing. The physical environment changes on many different time scales from seconds to million of years. There are daily and seasonal temperature cycles; ice ages are temperature cycles with a period of tens of thousands of years. Continental drift causes major climatic changes on even longer time scales. More subtle are the changes in the biological environment. Most creatures are both predator and prey; all, save perhaps viruses, are attacked by pathogens. These components of the biological environment are just as variable as the physical environment, perhaps even more so. Our picture of evolution should never be one of constant improvement in a static environment, but rather a desperate evolutionary race to avoid extinction in a constantly deteriorating environment. With this view, it is perfectly natural to assume that *medionigra* was more fit in the Oxford environment at some time prior to 1939 and less fit after 1939. Of course, we have no way to reconstruct history to find out if this is so. Nor can we rule out such possibilities *as a* sudden increase in the mutation rate due a transposable element or some other unorthodox mutational event.

The second type of selection occurs when there is overdominance ($h < 0$). In this case, the allele frequency approaches an equilibrium value, $p \rightarrow \hat{p}$, as is readily inferred from the graph of $\Delta_s p$ versus p illustrated in the left-hand. When p is close to zero, $\Delta_s p > 0$ and the allele frequency will increase. When p is close to one, $\Delta_s p < 0$ and p will decrease. Therefore, p must approach an equilibrium that is between zero and one. The equilibrium is that point where the allele frequency no longer changes, $\Delta_s p = 0$. From Equation 2, we see that this occurs when

$$\hat{p}h + (1-\hat{p})(1-h) = 0,$$

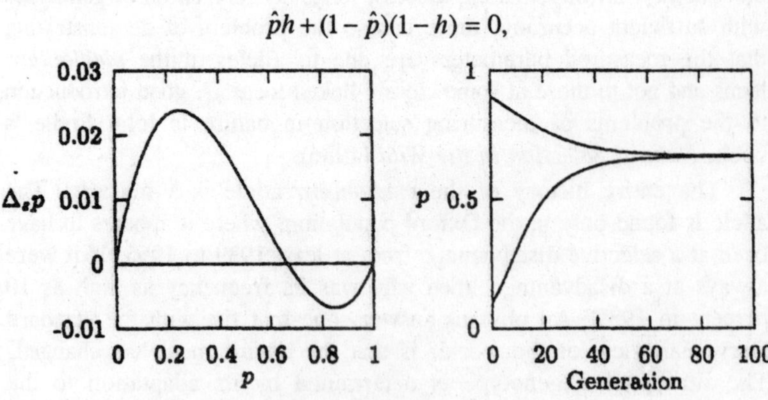

Fig. 10.4. Properties of balancing selection with h = – 0.5 and s = 0.1.

or

$$\hat{p} = \frac{h-1}{2h-1}. \qquad ...(4)$$

As both alleles are kept in the population in a balanced or stable equilibrium, this form of selection is called balancing selection.

As a technical aside, the method used to determine the existence of an internal equilibrium is called end-point analysis. This technique is very useful for describing the qualitative behaviour of complicated dynamical models when the global behaviour is too difficult to understand. End-point analysis could have been used for directional selection as well. There, the allele frequency increases when both rare and common, suggesting that $p = 1$ is a stable equilibrium and $p = 0$ is an unstable equilibrium.

The most thoroughly studied example of overdominance is the sickle-cell hemoglobin polymorphism found in many human populations in Africa. Hemoglobin, the oxygen-carrying red protein found in red blood cells, is a tetramer composed of two alpha chains and two beta chains. In native West and Central African populations, the *S* allele of beta hemoglobin reaches a frequency as high as 0.3 in some areas. The more common *A* allele is found at very high frequency in most other areas of the world. The two alleles differ only in that the *S* allele has a glutamic acid at its sixth amino position while the *A* allele has a vane. The glutamic acid causes the hemoglobin to form crystal aggregates under low partial pressures of oxygen, as occur, for example, in the capillaries. As a result, *SS* homozygotes suffer from sickle-cell anemia, a disease that is often fatal.

The *S* allele could not have reached a frequency of 0.3 unless *AS* heterozygotes are more fit than *AA* homozygotes. This is precisely the case in regions where malaria is endemic, for there the heterozygotes are somewhat resistant to severe forms of malaria. The resistance is due to the sickling phenomena, which makes red blood cells less suitable for *Plasmodium falciparum*. In an old study from 1961, it was shown that the viability of *AS* relative to *AA* is 1.176 in regions with malaria. Assuming that the fitness of *SS* is zero (s = 1), h = – 0.176. Plugging this into Equation 4 gives $\hat{p}$ = 0.87 or $\hat{q}$ = 0.13 for the *S* allele, which is nestled right in the middle of allele frequencies in regions with endemic malaria.

The complete story of selection on beta hemoglobin variation is more complicated than the bit presented here. A very readable, though somewhat dated, account may be found in *The Genetics of Human*

Populations by Lucca Cavalli-Sforza and Walter Bodmer (1971). Even our abbreviated account emphasizes once again that selection occurs in an environment that is always changing. The relevant environment for the sickle-cell polymorphism is the biological environment embodied in falciparum malaria. Populations of *Plasmodium* fluctuate in both time and space with consequent changes in the fitnesses of the beta hemoglobin genotypes. In areas without malaria, the form of selection shifts from balancing selection to directional selection.

The final form of selection, disruptive selection, occurs when there is under-dominance ($h > 1$). A graph of $\Delta_s p$ versus p in this case shows that p will decrease when rare and increase when near one. That's strange: The outcome of selection depends on the initial frequency of the allele! In fact, the allele frequency will approach zero if the initial value of p is less than $\hat{p}$, where $\hat{p}$ is given by Equation 4. The allele frequency will approach one if the initial value of p is greater than $\hat{p}$. If, by some bizarre chance, $p = \hat{p}$, the allele frequency will not change at all. $\hat{p}$ is an unstable equilibrium because the smallest change in p will cause the allele frequency to move away from $\hat{p}$. A small change in p might well be caused by genetic drift.

There are very few, if any, examples of underdominant alleles in high frequency in natural populations. However, the fact that closely related species sometimes have chromosomes that differ by inversions or translocations suggests that underdominant chromosomal mutations do occasionally cross the unstable equilibrium. The evolutionary forces that push the frequencies over the unstable point are not known, although both genetic drift and meiotic drive are likely candidates.

There is something unsatisfying about the description of the three forms of natural selection. They come off as a series of disconnected cases. One might have hoped for some unifying principle that would make all three cases appear as instances of some more general dynamic. In fact, Sewall Wright found unity when he wrote Formula 2 in the more provocative form

$$\Delta_s p = \frac{pq}{2\bar{w}} \frac{d\bar{w}}{dp}. \qquad \ldots(5)$$

(The symbol $d\bar{w}/dp$ is the derivative or slope of the mean fitness viewed as a function of the allele frequency p.) Equation 5 shows that $\Delta_s p$ is proportional to the slope of the mean fitness function. If the slope is positive, then so is $\Delta_s p$. As a result, selection will increase p and, because $d\bar{w}/dp > 0$, will increase the mean fitness of the population. If the slope is negative, p will decrease and, because $d\bar{w}/$

$dp < 0$, the mean fitness will increase once again. In other words, the allele frequency always changes in such a way that the mean fitness of the population increases. Moreover, the rate of change in p is proportional to the genetic variation in the population as measured by pq. Although we will not show it, the rate of change of the mean fitness, $\bar{w}$, is proportional to pq as well. Thus, selection always increases the mean fitness of the population and does so at a rate that is proportional to the genetic variation.

R. A. Fisher made a similar observation at about the same time as did Wright and called it the Fundamental Theorem of Natural Selection (Fisher 1958). Fisher showed that the change in the mean fitness is proportional to the additive genetic variation in fitness. As variances are always positive, the mean fitness will always increase when natural selection changes the allele frequency.

The Fundamental Theorem of Natural Selection is undeniably true for theoretical populations with simple selection at a single locus. However, with more loci or if fitness depends on the frequencies of genotypes or if it changes through time, the Fundamental Theorem no longer holds. Thus, it is neither fundamental nor a theorem; some have claimed that it has little to do with natural selection. Its biological significance has always been controversial. Yet, the metaphor suggested by the theorem that natural selection always moves populations upward on the "*adaptive landscape*" has proven to be a convenient one for simple descriptions of evolution without mathematics or deep understanding. The metaphor is stretched too far when applied to evolution for more than a few generations, as the change in fitness due to environmental change renders the metaphor inappropriate. Imagine climbing a mountain that keeps moving; despite your best efforts, the peak remains about the same distance ahead. That is the proper metaphor for evolution.

Mutation-Selection Balance

The vast majority of mutations of large effect are deleterious and incompletely dominant. They enter the population by mutation and are removed by directional selection. A balance is reached where the rate of introduction of mutations is exactly matched by their rate of loss due to selection. The equilibrium number of deleterious mutations is large enough to have a major effect on many evolutionary processes. Among these are the evolution of sex and recombination and the avoidance of inbreeding. Most of these mutations are partially recessive, $h < 1/2$, so their effects are not always apparent unless the population is made homozygous either by genetic drift or by inbreeding. In this section, we will study the balance between mutation and selection and then go on in the next section to describe the dominance relationships between naturally occurring alleles.

Following our labeling conventions, A_2 will represent the deleterious allele whose frequency is increased by mutation and decreased by directional selection. Selection will be assumed to be sufficiently strong so that the frequency of A_2 is very small. As a consequence, the most important effect of mutation is to convert A_1 alleles into A_2 alleles. The reverse happens as well, but has little influence on the dynamics and can be ignored. Suppose, therefore, that there is one-way mutation from allele A_1 to allele A_2,

$$A_1 \xrightarrow{u} A_2,$$

where u is the mutation rate, the probability that a mutation from A_1 to A_2 appears in a gamete.

The effects of mutation on p may be described in the same way as was done in the discussion of the balance between mutation and genetic drift. For an allele in the next generation to be A_1, it must have been A_1 in the current generation and it must not have mutated,

$$p' = p(1 - u).$$

The change in p in a single generation is

$$\Delta_u p = -up. \qquad \text{...(6)}$$

Mutation rates are usually very small: 10^{-5} for visible mutations at a typical locus in *Drosophila* to 10^{-9} for a typical nucleotide. Thus, the frequency of A_1 decreases very slowly while the frequency of A_2 increases very slowly. If selection against A_2 is sufficiently strong, it will keep the frequency of A_2 very low, allowing the approximation

$$\Delta_u p = -u + qu \approx -u, \qquad \text{...(7)}$$

because $q \approx 0$.

From Equation 2, we can write the change in the frequency of A_1 due to selection acting in isolation, when $q \approx 0$, as

$$\Delta_s p = \frac{pqs[ph + q(1-h)]}{1 - 2pqhs - q^2 s} \approx qhs. \qquad \text{...(8)}$$

The approximation is valid when $q \approx 0$, which implies that $p \approx 1$ and $\overline{w} \approx 1$.

At equilibrium, the change in the frequency of A_1 by mutation must balance the change due to selection,

$$0 = \Delta_u p + \Delta_s p$$
$$\approx -u + qhs,$$

which gives the equilibrium frequency of A_2,

$$\hat{q} \approx \frac{u}{hs} \qquad \text{...(9)}$$

The equilibrium frequency of a deleterious allele is approximately equal to the mutation rate to the allele divided by the selection against the allele in heterozygotes. Rare alleles *are* found mainly in heterozygotes, not homozygotes. Thus, it is not surprising that the equilibrium frequency of deleterious alleles depends on their fitness in heterozygotes rather than homozygotes.

Deleterious alleles cause problems for populations. One measure of these problems is the genetic load of the population,

$$L = \frac{w_{\max} - \bar{w}}{w_{\max}}, \qquad \text{...(10)}$$

where w_{max} is the fitness of the maximally fit genotype in the population. The closer the mean fitness of the population is to the fitness of the most fit genotype, the less is the genetic load.

The mean fitness of a population at equilibrium under the mutation-selection balance is

$$\begin{aligned}\bar{w} &= 1 - 2\hat{p}\hat{q}hs - \hat{q}^2 s \\ &\approx 1 - 2\hat{q}hs \\ &\approx 1 - 2u.\end{aligned}$$

Remarkably, the presence of deleterious mutations decreases the mean fitness by an amount, $2u$, that is independent of the strength of selection in heterozygotes. The genetic load in this case is simply

$$L = \frac{1 - (1 - 2u)}{1} = 2u \qquad \text{...(11)}$$

which follows from Equation 10 with $w_{max} = 1$. When selection is weak, the frequency of the deleterious allele will be higher, but the detrimental effect of each allele on the mean fitness of the population is slight. When selection is strong, the frequency of A_2 is less, but the effect is greater. Hence, the independence of the load on the strength of selection follows.

The biological significance of genetic load, like that of the Fundamental Theorem of Natural Selection, has been hotly debated over the years. Load is most useful when discussing deleterious alleles of measurable effect but is of dubious value for variation maintained by balancing selection or for directional selection of advantageous alleles.

Heterozygous Effects of Alleles

In 1960, Rayla Greenberg and James F. Crow published a landmark paper reporting the results of a study "undertaken in an attempt to determine whether the effects of recurrent mutation on the population and the deleterious effects of inbreeding are due primarily to a small number of genes of major effect or to the cumulative activity of a number of genes with individually small effects." The study did this and considerably more. Of particular interest was its suggestion that mutations of large effect are almost recessive ($s \approx 1$ and $h \approx 0$) while those of small effect are almost additive ($s \approx 0$ and $h \approx 1/2$). The prevailing view in the late 1950s was that most deleterious mutations are completely recessive ($h = 0$). This paper is not only historically and scientifically important, but also pedagogically valuable because it uses many of the ideas developed in this and the previous chapters. In addition, it introduces an experimental methodology that is central to population genetics.

The design of the experimental part of the Greenberg and Crow paper was developed in the late 1930s by Alfred Sturtevant and Theodosius Dobzhansky. Back then, most of the genetic variation affecting fitness was thought to be due to rare, recessive, deleterious alleles. As rare alleles are usually heterozygous, these mutations would not be expressed in wild-caught individuals. An obvious way to study this "hidden variation" is to make individuals homozygous, as this allows the expression of recessive mutations. *Drosophila* was the only suitable organism for such a study because it alone allowed the experimental manipulation of entire chromosomes on which recombination is suppressed.

The purpose of the design is to construct flies that are homozygous for their entire second chromosomes. The viabilities of these flies are then compared to those whose two second chromosomes are drawn independently from nature, thus mimicking random mating. The first class of flies will be called *inbred*, and the second class will be called *outbred*. The details of the design are as follows;

P_1 In the parental generation, $+_n/+'_n$ represents represents one male fly, obtained from nature or from an experimental population, that initiates the nth line of crosses. The symbols $+_n$ and $+'_n$, stand for the two second chromosomes found in the male fly. One of the second chromosomes of this fly will ultimately be made homozygous. As there is no recombination in male *Drosophila*, the chromosomes in this original male

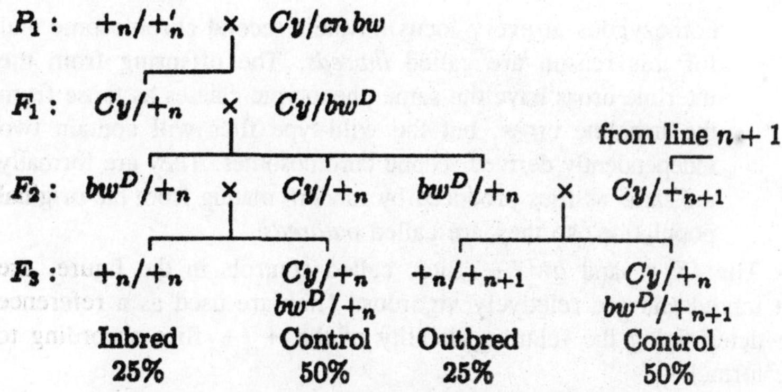

Fig. 10.5. The Drosophila melanogaster crosses used to uncover hidden variation. In each cross, the male is on the left.

remain intact. The male is crossed to *a Cy/cm bw* female. *Cy* is a dominant second chromosome mutation, Curly wing, that is placed on a chromosome with one paracentric inversion on each arm to block recombination. The other chromosome has two recessive mutations, cinnabar eyes *(cn)* and brown eyes (*bw*). This initial cross is repeated 465 times, each repetition using an independently obtained male. Each repetition is called a line; the lines are numbered sequentially from 1 to 465.

F_1 A single $Cy/+_n$ male from each line is crossed to *a* Cy/bw^D female. The female *Cy* chromosome in this step is a slightly fancier version of the previous *Cy* chromosome, with a pericentric inversion, *SM*1, providing extra safeguards against recombination. The homolog to *Cy* in the female contains a dominant brown-eye mutation, bw^D. This is the critical step in the design as it assures that only a single wild-caught second chromosome is used.

F_2 Two different crosses occur in this generation. The first is a mating of a a $bw^D/+_n$ male to a $Cy/+_n$ female from the same line (a brother-sister mating). The second is a cross of a $bw^D/+_n$ male to a $Cy/+_{n+1}$ female from the (n + 1)st line.

F_3 The offspring from the brother-sister F_2 cross will fall into four classes: $+_n/+_n$, $Cy/+_n$, $bw^D/+_n$, and Cy/bw^D, which are easily recognized because *Cy* and bw^D are both dominant mutations. According to Mendel's law of segregation, these four classes should be equally frequent. However, as the Cy/bw^D flies are not used in the analysis. The $+_n/+_n$ flies are homozygous at every locus on their second chromosome and for this reason are called *inbreds*. The offspring from the interline cross have the same phenotypic classes as those from the intraline cross, but the wild-type flies will contain two independently derived second chromosomes. They are formally the same as flies produced by random mating from the original population, so they are called *outbreds*.

The $Cy/+_n$ and $bw^D/+_n$ flies, called controls in the figure, are not inbred and are relatively vigorous. They are used as a reference for determining the relative viability of the $+_n/+_n$ flies according to the formula

$$+_n/+_n \text{ viability} = \frac{2 \times \text{number of } +_n/+_n \text{ flies}}{\text{number of } Cy/+_n \text{ and } bw^D/+_n \text{ flies}}.$$

The determination of the viability of outbred flies is done in a similar fashion.

Viability class 0 is made up of all lines with relative viability in the interval $0 \leq v < 0.1$, class 0.1 includes those in the range $0.1 \leq v < 0.2$, and so forth. The most striking aspect of the figure is the bimodal distribution of viabilities in the inbred flies. The left mode is due to "recessive" lethal mutations found on 106 of the 465 chromosomes examined. In other words, about 23 percent of all second chromosomes carry at least one mutation that is lethal in the homozygous state. As a single second chromosome represents about 20 percent of the total genes in *Drosophila*, it follows that most individuals carry at least one

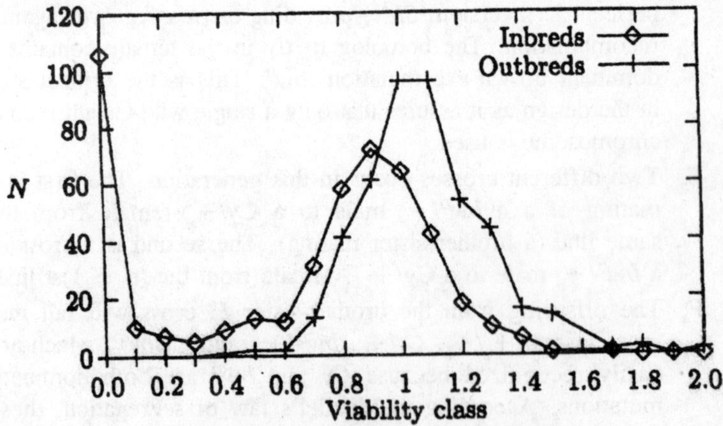

Fig. 10.6. The number of lines in each viability class among the inbred and outbred flies in the Greenberg and Crow experiment.

lethal mutation. *Drosophila* is not unusual in this regard. Most diploid organisms, including ourselves, carry a similar number of lethal mutations.

The graphed viabilities of inbred flies dips into a deep valley separating a left-hand lethal peak from the much broader right-hand deleterious peak. The mode of the deleterious peak lies a bit to the left of the mode of the outbred flies. Thus, flies homozygous for their second chromosomes do not fare as well as outbred flies, even if their second chromosomes are free of lethal mutations. Being homozygous at all loci is clearly bad; this is an example of a more general phenomenon called *inbreeding depression*, which could be defined as the reduction in mean fitness due to increased homozygosity.

At this point, it might seem that we have in hand some evidence on the value of the heterozygous effect, h, for the hidden variation uncovered in this experiment. Perhaps the observation that inbreds are less viable than outbreds implies that overdominance is the prevalent form of dominance. While plausible, inbreeding depression does not allow this conclusion, as may be shown by considering the contribution of a single second-chromosome locus to viability. The fitnesses and frequencies of the genotypes at this locus in inbred and outbred flies of the Greenberg and Crow experiment are as follows:

Genotype:	A_1A_1	A_1A_2	A_2A_2
Relative fitness:	1	$1 - hs$	$1 - s$
Inbred frequencies:	p	0	q
Outbred frequencies:	p^2	$2pq$	q^2

The experiment tells us that the mean fitness of the inbred flies is less than that of the outbred flies. What does this say about the values of s and h in the table?

The mean viability of the inbred flies is

$$(p \times 1) + [q \times (1 - s)] = 1 - qs.$$

The mean viability of the outbred flies is just the mean fitness of a randomly mating population, as given in Equation 3. If the outbreds are more viable than the inbreds,

$$1 - 2pqhs - q^2s > 1 - qs \qquad ...(12)$$

In sequence, cancel the ones, cancel qs, move q to the right side of the inequality, and clean up to get

$$h < 1/2 \qquad ...(13)$$

Thus, inbreeding depression implies that h is less than one-half. Recalling that in the s-h parameter system the A_1A_1 genotype is always more fit than A_2A_2, we can conclude that inbreeding depression implies that the fitness of the heterozygote is, on average, closer to that of the more fit homozygote. Recessive deleterious mutations have this property, as do overdominant mutations. While intuition may have suggested that inbreeding depression implies overdominance, we now see that it only limits the heterozygous effect to being less than one-half. Of course, this is a tremendous step forward in our quest to learn more about the alleles responsible for genetic variation in fitness. Already we can pay less attention to underdominant and recessive advantageous mutation and focus on those mutations with $h < 1/2$.

The next step in Greenberg and Crow's analysis is an indirect but brilliant inference about the relationship between h and s, which begins with some formulae that relate the mean relative viabilities of flies to the frequencies and effects of deleterious mutations. Three viability estimates are required:

A The average relative viability of outbred flies. $A = 1.008$.

B The average relative viability of inbred flies: $B = 0.632$. When other studies are included, B is found to lie between 0.614 and 0.656.

C The average relative viability of inbred flies without lethal mutations on their second chromosomes: $C = 0.842$. Among all such studies, C ranges from 0.829 to 0.860.

Each second chromosome is imagined to have an unknown number, n, of loci that are capable of mutating to deleterious and lethal alleles. The frequency of the deleterious allele at the ith locus is called q_i and its selection coefficient, which is thought to be small, s_i. Similarly, the frequency of the lethal mutation at the ith locus is called Q_i and its selection coefficient, which is close to one, S_i. Inbred flies have lower viabilities than outbred flies because they are more likely to carry one or more of these mutations in the homozygous state. The probability that a particular inbred fly is homozygous for a deleterious

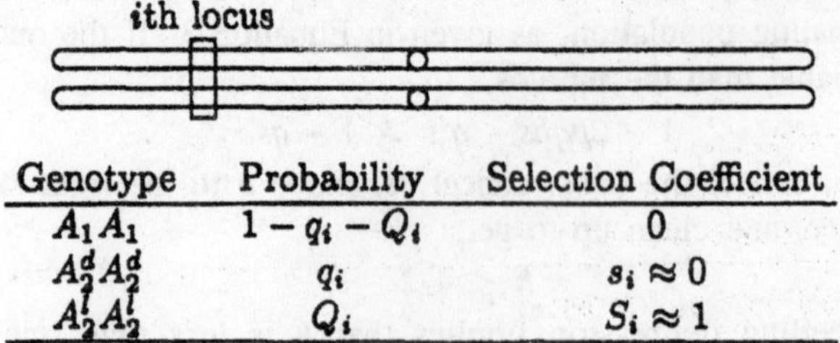

Genotype	Probability	Selection Coefficient
A_1A_1	$1 - q_i - Q_i$	0
$A_2^dA_2^d$	q_i	$s_i \approx 0$
$A_2^lA_2^l$	Q_i	$S_i \approx 1$

Fig. 10.7. The possible states of a typical locus in an inbred fly.

allele at the i locus is q_i. The probability that it dies from this allele, given that it is homozygous, is s_i. The probability that it is both homozygous and dies is $q_i s_i$. Finally, the probability that it survives is $1 - q_i s_i$.

If—and this is a big 'if'—the loci act independently in their effects on the probability of survival, the probability that a particular inbred fly survives to adulthood is

$$B = A\prod_{i=1}^{n}(1-q_i s_i)(1-Q_i S_i). \qquad ...(14)$$

The factor A represents the probability of survival for an outbred fly. Inbreds may die for all of the various reasons that outbreds may die, plus some more, which are captured in the product term. Using Equation 14 may be approximated by

$$B = Ae^{-D-L}, \qquad ...(15)$$

where $D = \Sigma q_i s_i$ is called the *detrimental load* and $L = \Sigma Q_i S_i$ is called the *lethal load*. These loads differ from the genetic load described in the previous section in that they describe the fitness reduction in inbred rather than outbred flies.

By taking the natural logarithm of both sides of Equation 3.15 and rearranging, we obtain

$$\begin{aligned} D + L &= \ln(A) - \ln(B) \\ &= \ln(1.008) - \ln(0.632) \\ &= 0.4668, \qquad ...(16) \end{aligned}$$

which relates the estimates of relative viabilities to the parameters of deleterious alleles. D itself may be obtained from

$$\begin{aligned} C &= A\prod_{i=1}^{n}(1-q_i s_i) \\ &\approx A\exp(\sum_i q_i s_i) \\ &= Ae^{-D} \end{aligned}$$

by taking logarithms of both sides and rearranging to obtain

$$D = \ln(A) - \ln(C) = 0.1799.$$

This result may be combined with Equation 16 to get

$$L = \ln(C) - \ln(B) = 0.2868.$$

The final quantity of interest, the $D : L$ ratio, is

$$\frac{D}{L} = \frac{\Sigma q_i s_i}{\Sigma Q_i S_i} = 0.627. \qquad ...(17)$$

This concludes the association of estimated values from the experiment to parameters from the population. The next task is to interpret the result.

In the late 1950s, most population geneticists believed that the majority of mutations in natural populations are deleterious with similar, and small, heterozygous effects. Greenberg and Crow claimed that this view was not compatible with a $D : L$ ratio of 0.627 by the following argument. Using Formula 9 for the equilibrium values of q_i and Q_i and canceling the selection coefficients, Equation 17 becomes

$$\frac{D}{L} = \frac{\Sigma u_i / h_i}{\Sigma U_i / H_i} \approx \frac{\Sigma u_i}{\Sigma U_i},$$

where u_i and U_i are the mutation rates to deleterious and lethal mutations at the ith locus, respectively. The hypothesis of similar heterozygous effects, $h \approx h_i \approx H_i$, allows the final cancelation of the heterozygous effects. Under this hypothesis, the $D : L$ ratio is equal to the ratio of the total deleterious mutation rate to the total lethal mutation rate. These rates may be estimated in the laboratory and their ratio, circa 1960, was known to be between 2 and 3. Spontaneous deleterious mutations are two to three times more likely to have a small effect than to be lethal. Thus, the hypothesis of equal heterozygous effects must be rejected because of the small value of the $D : L$ ratio.

Greenberg and Crow suggested an alternative hypothesis: Suppose there is an inverse relationship between h and s. For example, suppose hs is constant across mutations. In this case,

$$\frac{D}{L} = \frac{\Sigma (u_i / h_i s_i) s_i}{\Sigma (U_i / H_i S_i) S_i} = \frac{\Sigma u_i s_i}{\Sigma U_i S_i},$$

because, by assumption, $h_i s_i \approx H_i S_i$. Now the $D : L$ ratio is the ratio of quantities that are sums of selection coefficients weighted by their mutation rates. These quantities may also be estimated in the laboratory, and the ratio turns out to be 0.711, remarkably close to 0.627, thus giving support to the hypothesis of an inverse homozygous-heterozygous effect.

Greenberg and Crow go on to consider a few other hypotheses but finally return to this one and with it their suggestion that there is an "inverse heterozygous-homozygous effect" for deleterious mutations uncovered by inbreeding. Mutations of large effect, like lethals, are almost recessive ($h \approx 0$); alleles of small effect have a much greater heterozygous effect, perhaps close to one-half, but not greater than one-half because of inbreeding depression.

The inverse heterozygous-homozygous effect is not an isolated phenomenon. Rather, it is an instance of a more general framework whose origins may be traced to the 1920s and the first efforts to understand the dominance relationships between alleles affecting phenotypes. Some visible phenotypes also exhibit an inverse homozygous-heterozygous effect, the most celebrated being the white-eye mutants in *Drosophila*.

Sewall Wright, among others, created a model to explain the inverse effect that used well-established properties of enzyme pathways. The model is of a locus whose product, an enzyme, catalyzes one step in a critical enzyme pathway. The horizontal axis is the activity of the enzyme in units chosen so that an activity of one is "normal." The vertical axis is the fitness of the genotype as a function of its enzyme activity. (In the original formulation the vertical axis was a quantitative measure of the phenotype.) If there is no activity, the pathway does not function and the fitness is zero. As the activity increases, there is a rapid increase in fitness because the pathway can now produce its product. With further increases in activity, the pathway begins to function normally and the augmentation of fitness decreases. There is a "law of diminishing returns" as enzyme activity approaches normal levels. It is fairly well established that most enzymes have such high activities that a reduction in activity of one-half has only a small effect on the functioning of the pathway.

In Wright's model, a lethal mutation is one that produces a defective enzyme with no activity. In the context of our two-allele

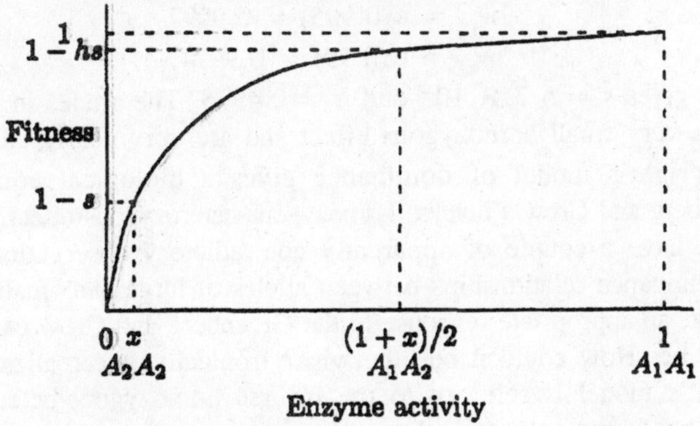

Fig. 10.8. Sewall Wright's model of dominance applied to viability.

model, the A_2A_2 homozygote has no activity and dies. The A_1A_1 homozygote produces normal enzymes and has a scaled activity of one. The activity of the A_1A_2 heterozygote is one-half that of a normal homozygote because it contains one allele making a normal enzyme and one making a defective enzyme. Its scaled activity is 1/2, yet its fitness is close to that of A_1A_1. In this case, we have $s = 1$ and h close to zero.

A deleterious mutation of small effect will be one whose enzyme activity is only slightly less than the normal activity. Again, the heterozygote's activity will be halfway between the normal activity and the mutant homozygote activity. The fitness of the heterozygote, however, will be only slightly less than halfway between that of the two homozygotes because of the concave form of the fitness function. In this case, we have s close to zero and h close to one-half. Hence, the inverse homozygous-heterozygous effect of the Greenberg-Crow experiment.

For example, the function is

$$w(x) = \frac{(1+a)x}{a+x},$$

with $a = 0.06$. When A_2 is a lethal mutation, we have

$$w_{11} = w(1) = 1$$
$$w_{12} = w(1/2) = 0.946$$
$$w_{22} = w(0) = 0.$$

From these we get $s = 1$ and $h = 0.053$, which represent a high degree of dominance for the A_1 allele. If A_2 is only slightly deleterious, with an enzyme activity of 0.99, then

$$w_{11}\, w(1) = 1$$
$$w_{12} = w(0.995) = 0.9997$$
$$w_{22} = w(0.99) = 0.9994,$$

which gives $s = 5.7 \times 10^{-4}$ and $h = 0.4976$. The alleles in this case have a very small homozygous effect and are very nearly additive.

Wright's model of dominance gives a biological context for Greenberg and Crow's inverse homozygous-heterozygous effect. Without it, we have a couple of apparently contradictory observations about the dominance relationships between alleles of large and small effects. With it, an appropriate response to the Greenberg and Crow experiment would be: How could it be otherwise? Ironically, the realization that Wright's model is relevant to the inverse homozygous-heterozygous effect has come only recently.

There is no room in our functional view of dominance for overdominance. Wright's model, in particular, allows only incomplete dominance. The compelling biological appeal of his model suggests that overdominance, when it exists, must be due to unusual circumstances. The sickle-cell polymorphism is an example: Overdominance results from the interaction of the solubility of a globin mutation and the growth requirements of a protozoan. This very peculiar situation would probably not generalize to a sizeable fraction of loci.

The analysis of the Greenberg and Crow experimental results may strike some as unacceptably abstract because it gives no examples of deleterious alleles of small effect. Of course, in 1960 there were no molecular techniques to allow this. Today, such techniques are available and so are some examples of mutations that could plausibly contribute to Greenberg and Crow's detrimental class. Null alleles, alleles at enzyme-encoding loci that lack measurable activity, are the best-known group of putative deleterious mutations. In a large study of null alleles in *Drosophila* populations, Chuck Langley, Bob Voelker, and their colleagues estimated that the average frequency of null alleles is $\bar{q} = 0.0025$. Previously, Terumi Mukai and Clark Cockerham had estimated the average mutation rate to null alleles to be $\bar{u} = 3.86 \times 10^{-6}$. If the nulls are deleterious and the population is at equilibrium, Equation 9 gives

$$\overline{hs} = u/q = 1.5 \times 10^{-3}.$$

Surprisingly, all but 1 of the the 20 autosomal null alleles showed no obvious deleterious effects when homozygous in the laboratory, so these 19 do not contribute to the lethal class of the Greenberg and Crow experiment. No effort was made to estimate the relative fitnesses of the nulls as homozygotes or heterozygotes, so we cannot say what their heterozygous effects are. These estimates will probably never be made because measuring fitness differences as small as 10^{-3} is impractical, if not impossible.

The main insights of the Greenberg-Crow experiments have withstood the tests of time remarkably well. In a monumental experiment, Terumi Mukai measured viability effects of newly arising mutations and directly confirmed the existence of the inverse homozygous-heterozygous effect. Mukai also looked at viability variation in natural populations and found no evidence for overdominance, thus supporting our view that overdominance is not a common phenomenon. The ratio of deleterious to lethal mutation rates is now thought to be closer to 10 than to 2 to 3 as used by Greenberg and Crow. The paper by Mike Simmons and James Crow (1977) contains a good summary of these more recent results.

Changing Environments

Greenberg and Crow and subsequent workers have shown that alleles with very small effects on viability are close to additive. Many of these alleles in natural populations are undoubtedly maintained by mutation-selection balance; others could be maintained by balancing selection. However, our theoretical investigations of natural selection showed that balancing selection occurs only when there is overdominance, $h < 0$. If all of the experimental work, except for a few examples of strong selection like the sickle-cell polymorphism, argues against overdominance, it would be rather silly to suggest that balancing selection is an important contributor to genetic variation in fitness. Or would it? Perhaps there are situations where balancing selection can occur without overdominance.

Two of our examples of selection suggest such a situation, that of selection in a changing environment. In *Panaxia*, the fitness of the *medionigra* allele must have changed from advantageous to disadvantageous at least once in the recent past. In addition to this temporal variation in fitness, there must also be spatial variation in fitness, as medionigra was in moderate frequency only in the Oxford population. Similarly, the overdominance of the sickle-cell allele occurs only in areas with high levels of malaria. In other areas, there is incomplete dominance.

As instructive as these examples are, the idea that the fitness of a genotype depends on the environment is so obviously true that little in the way of support needs to be mustered. If fitnesses do depend on the state of the environment, as they surely must, then they must just as assuredly change in both time and space, driven by temporal and spatial fluctuations in the environment.

Can changing fitnesses in time and space result in balancing selection without overdominance? The answer is yes, but the route to the answer is one of the more difficult proofs in theoretical population genetics. While the proof is difficult, the result is rather intuitive. Imagine that the fitnesses of the A_1A_1 and A_2A_2 homozygotes can be written as $w_{11} = 1 + s_i$ and $w_{22} = 1 - s_i$, where s_i is a selection coefficient whose value depends on the state of the environment in the ith habitat. Imagine further that the fitness of the heterozygote is one, which corresponds to $h = 1/2$. If the environment changes such that s_i is positive in some habitats and negative in others, then a balanced polymorphism is possible if the absolute value of the mean of s_i across habitats is less than the variance in s_i,

$$|E\{s_i\}| < \text{Var}\{s_i\}.$$

That is, if the variance in fitness across habitats is large enough to overcome any mean advantage one allele may have over another, then a balanced polymorphism will occur without overdominance.

Although the general proof of this result is quite difficult, the core of the argument is actually rather simple and will be illustrated here with a very particular example. The example is of spatial variation in fitness. We imagine a simple pattern where, in some parts of a species' range, the A_1 allele is favoured over the A_2 allele, and in other parts the opposite. Were there no migration, it is clear that the A_1 allele would fix in those places where it is favoured, the A_2 allele would fix where it is favoured, and the species would be polymorphic because of balancing selection without overdominance. We could end here with a satisfied grin except for one awkwardness: real species migrate. Migration tends to make species genetically more uniform across their range and can promote the fixation of an allele that is more fit on average than all others across the entire range of the species. Obviously, we have some work to do; we must investigate the conditions that lead to polymorphism when migration occurs between subpopulations.

There are n subdivisions in the environment, called *patches*, each of relative size c_i, $\Sigma c_i = 1$. In each generation, after selection occurs within a patch, a fraction m of alleles from each patch are exchanged at random with alleles from the other patches. The probability of an exchange with a particular patch is proportional to the relative size of that patch, c_i. The fitnesses of the three genotypes at the A locus vary

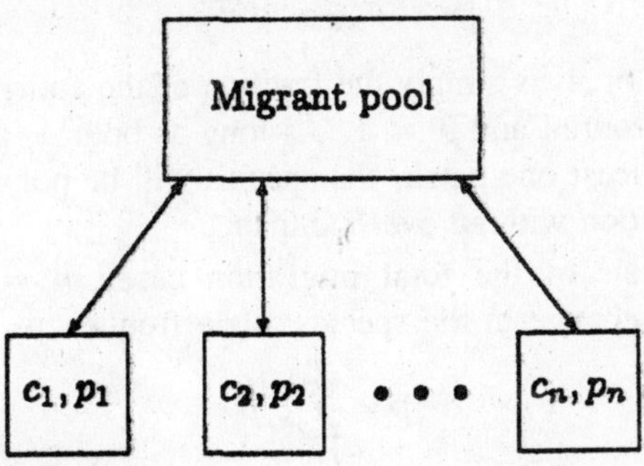

Fig. 10.9. A model of selection in a species with subpopulations, each of different relative size, c_i, and fitness.

across patches. The change in p_i, the frequency of the A_1 allele in the ith patch, will be written $\Delta_s p_i$. The frequency of A_1 in the ith patch after a round of selection and migration is

$$p_i' = (1-m)(p_i + \Delta_s p_i) + m\sum_{j=1}^{n} c_j(p_j + \Delta_s p_j). \qquad ...(18)$$

An allele within a patch is from a resident with probability 1 – m and from an immigrant with probability m. If it is from an immigrant, then the probability that it is A_1 is just the frequency of A_1 in the entire species after selection. It is as if during each generation all of the subpopulations contribute to a "migrant pool" from which they then choose alleles to replace those that emigrated. Notice that alleles rather than genotypes migrate. We obviously lose some biological realism with this sort of migration, but we gain a great deal in the simplicity of the model. Nothing that we say would change significantly were we to use a model with individuals migrating.

Equation 18 cannot be analyzed in the same way as the other models of selection because it is much too complex. However, by examining some special cases we will find out everything we need to know. The first specialization is to one of two extreme migration rates, no migration, $m = 0$, or total migration, $m = 1$; the second is to incomplete dominance, $0 < h < 1$.

For the no-migration case, $m = 0$, the behaviour of the model is simple: fixation will occur in each patch for the favoured allele. For those patches where $w_{11} > w_{22}$, $p_i = 1$, otherwise $p_i = 0$. As the frequency of A_1 for the species is the average allele frequency across subpopulations,

$$p = \sum_{i=1}^{n} c_i p_i.$$

The frequency of A_1 is simply the fraction of the entire population for which A_1 is favoured and $p_i = 1$. As long as both A_1 and A_2 are each favoured in at least one patch, the species will be polymorphic due to balancing selection without overdominance.

The analysis of the total migration case, $m = 1$, begins by examining the change in the species allele frequency,

$$p = \sum_{i=1}^{n} c_i p_i.$$

Multiply both sides of Equation 18 by c_i and add the n equations to get

$$\sum_i c_i p_i' = (1-m)\sum_i c_i(p_i + \Delta_s p_i) + m\sum_i c_i \sum_{j=1}^{n} c_j(p_j + \Delta_s p_j)$$

or

$$p' = p + \overline{\Delta_s p},$$

where

$$\overline{\Delta_s p} = \sum_{j=1}^{n} c_j \Delta_s p_j$$

is the average of the changes in p_i across patches. The change in p in a single generation becomes

$$\Delta_s p = \overline{\Delta_s p}. \qquad ...(19)$$

On the surface, it looks as if we just made a difficult problem almost trivial. Equation 3.19 says that the change in the species allele frequency is the average of the changes of allele frequencies in the subpopulations. And the migration rate has totally disappeared! However, we can't know $\overline{\Delta_s p}$ without knowing all of the p_i, and they do depend on the migration rate, as seen in Equation 18.

When there is complete migration, $m = 1$, then Equation 19 does contain all of the information because the allele frequencies in all of the patches are equal at the start of each round of selection. Using Equation 1, Equation 19 becomes

$$\Delta_s p = pq\sum_{i=1}^{n} c_i \frac{p(w_{11}^i - w_{12}^i) + q(w_{12}^i - w_{22}^i)}{p^2 w_{11}^i + 2pqw_{12}^i + q^2 w_{22}^i},$$

where the superscript i on the fitnesses indicates that the fitness depends on conditions in the ith patch. Rather than plowing directly into this equation, first specialize to the case of additive alleles ($h = 1/2$) and introduce some symmetry by setting $w_{11}^i = 1 + s_i$, $w_{12}^i = 1$, and $w_{22}^i = 1 - s_i$, giving

$$\Delta_s p = pq\sum_{i=1}^{n} \frac{c_i s_i}{1 + s_i(p-q)}. \qquad ...(20)$$

(In deriving this equation, use $p^2 - q^2 = p - q$.)

All that we need from Equation 3.20 is the answer to our question: Does balancing selection occur with incomplete dominance? The easiest route to the answer is to use end-point analysis as we did in the overdominance case. First, does p increase when small? When $p \approx 0$, the sign of Equation 20 is determined by

$$\sum_{i=1}^{n} c_i \frac{s_i}{1-s_i} \approx \sum_{i=1}^{n} c_i (s_i + s_i^2),$$

where the approximation comes from Equation 7. A_1 will increase when rare ($p \approx 0$) if

$$\sum_{i=1}^{n} c_i (s_i + s_i^2) > 0,$$

which is the same as

$$-\sum_{i=1}^{n} c_i s_i < \sum_{i=1}^{n} c_i s_i^2.$$

The condition that A_2 increases when rare is obtained in the same way and is

$$\sum_{i=1}^{n} c_i s_i < \sum_{i=1}^{n} c_i s_i^2.$$

Both conditions will be met when

$$\left| \sum_{i=1}^{n} c_i s_i \right| < \sum_{i=1}^{n} c_i s_i^2, \qquad \text{...(21)}$$

which is a sufficient condition for a polymorphism. (We have used the argument that, if a and b are both positive and if both a $< b$ and $-a < b$, then necessarily $|a| < b$.) If, for example, the absolute value of s_i is the same in each patch and if A_1 is favoured in precisely one-half of the patches, then the left side of the polymorphism condition is zero and the right side is greater than zero; hence, polymorphism will occur without overdominance.

As written, Equation 21 is not particularly informative. It can be made more so by recognizing that the left side is the absolute value of the average selective advantage of the A_1 allele. If the s_i are viewed as random quantities in a species with a large number of patches may be used to write the average as the expectation of s_i, $E\{s_i\}$. The right side may be approximated with the variance in s_i. From Equation 2, we know that

$$E\{s_i^2\} = \text{Var}\{s_i\} + E\{s_i\}^2.$$

If the mean and the variance of s_i are both very small and of similar orders of magnitude, then the square of the mean will be much smaller than the variance, $E\{s_i\}^2 \ll \text{Var}\ \{s_i\}$, so

$$E\{s_i^2\} \approx \text{Var}\{s_i^2\}.$$

Now we can rewrite Equation 21 in the much more suggestive form,

$$|E\{s_i\}| < \text{Var}\{s_i\} \qquad ...(22)$$

If the magnitude of the average selective advantage of an allele is less than the variance in fitness, polymorphism will occur. Said another way, if the variance in fitness is great enough to overcome the average selective advantage of alleles, polymorphism will occur. The more variable the environment, the more polymorphism.

When more complicated models of selection with incomplete dominance in a random environment are examined, the conditions for polymorphism are almost always in a form similar to Equation 22. Typically, there is a coefficient on the right hand side that reflects the particular mixture of temporal and spatial components of the fluctuations and the dominance relationships between alleles.

The condition for polymorphism will never be met unless some of the s_i are positive and some negative. This was the only condition required for polymorphism in the no-migration case, $m = 0$. While this condition is necessary for polymorphism when $m = 0$ and $m = 1$, it is sufficient only when $m = 0$. Migration generally makes polymorphism less likely when there is incomplete dominance because migration averages out the very environmental fluctuations that maintain the variation.

Given the ease with which fluctuating environments maintain variation and the fact that temporal fluctuations can cause the fixation of alleles, it is not surprising that the main alternative to the neutral theory as an explanation for molecular evolution and polymorphism is based on selection in a random environment. With fluctuating environments, there is balancing selection in nature, yet in the laboratory the experimentalist will see incomplete dominance. It could be that most of the variation that Crow and Greenberg called deleterious is maintained by fluctuating environments and should be called something else. At the time of this writing, there is no way to know whether the "deleterious" load is due mainly to alleles held in the population by mutation-selection balance or by balancing selection.

Selection and Drift

Our discussion of directional selection left the impression that the most fit allele eventually reaches a frequency of one. This is true for alleles of moderate frequency but is definitely not true for alleles with only one or a few copies in the population. These alleles are subject to the vagaries of Mendel's law of segregation and to demographic stochasticity. It is easy to see that the fate of a single copy of an allele with, say, a 1 percent advantage is determined mostly by chance. If its frequency should become moderate, then its average selective advantage can overcome the effects of genetic drift.

The interaction of drift and selection is more complex than that of mutation and drift because the strength of selection changes with the frequency of the allele. (Recall the factor pq in Δp.) Natural selection becomes a very weak force for rare alleles, as weak *as* or weaker than genetic drift when only a few copies of the allele are in the population. For example, there is only one copy of a new mutation, so its frequency when it first enters the population is $1/(2N)$. The strength of selection in this case is roughly $1/(2N)$ times s, which is less than the strength of genetic drift. When the frequency of the allele becomes larger, then the strength of selection is determined mainly by the selection coefficient, s. If $s \gg 1/(2N)$, selection will dominate drift for common alleles. Except for cases where $s \approx 1/(2N)$, selection and drift interact only in the dynamics of rare alleles. We will now describe this interaction.

In a finite population, a new advantageous mutation is usually lost because of genetic drift. This surprising result comes from the formula for the probability of ultimate fixation of the A_1 allele given its initial frequency,

$$\pi_1(p) = \frac{1 - e^{-2Nsp}}{1 - e^{-2Ns}}, \qquad \text{...(23)}$$

which applies to the case $h = 1/2$. (The derivation of Equation 23 will be postponed.) The subscript on $\pi_1(p)$ is a reminder that the function refers to the fixation probability of the A_1 allele.

The most important application of Equation 23 is for the fixation probability of a new mutation, $p = 1/(2N)$, which is

$$\pi_1\left(\frac{1}{2N}\right) = \frac{1 - e^{-s}}{1 - e^{-2Ns}}. \qquad \text{...(24)}$$

If s is so small that $e^{-s} \approx 1 - s$ and if $2Ns$ is so large that $e^{-2Ns} \approx 0$, then $\pi_1 \approx s$. Equation 23 is for $h = 1/2$, in which case the selection coefficient of the heterozygote is $s/2$. Thus, the fixation probability is twice the selective advantage of the heterozygote. This result holds more generally,

$$\pi_1 \approx 2(1 - h)s. \qquad \text{...(25)}$$

While proof of this assertion for arbitrary dominance will not be given, it is in accord with the general observation that selection on rare alleles depends on the fitness of the heterozygote.

Equation 25 tells us, for example, that a new mutation with a 1 percent advantage when heterozygous, $hs = 0.01$, has only a 2 percent chance of ultimately fixing in the population. A 1 percent advantage represents rather strong selection. In a very large population, say $N = 10^6$, 1 percent selection will overwhelm drift once the allele is at all common. Yet, 98 percent of such strongly selected mutations are lost. Think of all the great mutations that failed to get by the quagmire of rareness!

There is a much more important implication of our result. Imagine what happens when a species is challenged by a change in its environment. There may be mutations at many loci which, if fixed, will handle the new environment with a new adaptation. Which of these mutations gets there first, as it were, is determined to a great extent by chance. Adaptive evolution is, by its very nature, random. Were we to construct two perfect replicates of Earth, each with exactly the same species and sequences of environments, the course of evolution would be different in the two replicates. Evolution is not repeatable.

It is also possible for a deleterious mutation to fix in the population. The probability of fixation of the A_2 allele, given an initial frequency q, is given by

$$\pi_1(q) = 1 - \pi_1(1-q) = \frac{e^{2Nsq} - 1}{e^{2Ns} - 1}.$$

For a new mutation, $q = 1/(2N)$, and small s we have

$$\pi_2\left(\frac{1}{2N}\right) \approx \frac{s}{e^{2Ns} - 1}. \qquad \text{...(26)}$$

In very large populations, large enough that $2Ns \gg 1$, the probability of fixation of a deleterious allele becomes very small. However, when $2Ns$ is close to or less than one, the fixation of deleterious alleles can occur with reasonable probability. An unexpected application of this observation is in molecular evolution.

We have already disucsed that the rate of substitution of neutral alleles, k, is the mean number of mutations entering the population each generation, $2Nu$, times the probability of fixation of any one of them, $1/(2N)$. For selected alleles, we need only use π_i $(1/(2N))$ for the fixation probability to obtain

$$k = 2Nu\pi_i(1/2N). \qquad ...(27)$$

For advantageous mutations, Equation 25 shows that

$$k = 4Nu(1 - h)s.$$

When compared to the rate of substitution of neutral alleles, $k = u$, we see that selected molecular evolution depends on all three parameters of the model, rather than on the mutation rate alone. Many people have used this contrast to argue that selected molecular evolution is not compatible with the molecular clock but is compatible with neutral evolution.

It is also possible that most of molecular evolution involves the fixation of deleterious alleles. In this case, using Equation 26 in Equation 27, we have

$$k = \frac{2Nus}{e^{2Ns} - 1}. \qquad ...(28)$$

At first blush, this seems like a silly idea. How could most of the amino acid substitutions in proteins be deleterious? Certainly, evolution is not lowering the mean fitness of the population! In fact, this idea has a lot to recommend it. For example, it can explain three observations about molecular evolution that the neutral theory cannot explain. The first is the observation that the rate of substitution of amino acids in proteins is lower than the rate of substitution of nucleotides in noncoding regions. If most amino acid mutations are deleterious and most nucleotide substitutions in noncoding regions are neutral, then the amino acid rate should be lower.

The second observation is that there is only a slight generation-time effect in protein evolution but a pronounced generation-time effect in silent (noncoding) evolution. Recall that the generation-time effect is the observation that creatures with shorter generation times evolve faster than do those with longer generation times. The neutral theory predicts a generation-time effect if the mutation rate per generation is fairly similar across species, as it is thought to be. Mice, for example, should exhibit a rate of substitution that is much higher than that of elephants. However, if most mutations are deleterious, then the probability of fixation of these mutations is lower in mice than elephants because the population size of mice is much larger than that of elephants. The consequent lower rate of substitution in mice cancels the generation-time effect.

The third observation is the narrow range of heterozygosities across groups of organisms that are thought to have very different population sizes. We argued that the narrow range could be due to the effective population sizes of species being more similar than their current actual sizes due to fluctuations in their population sizes. Another explanation involves the idea of an effective neutral mutation rate, u_e. If most mutations are deleterious, then only those with selection coefficients close to $1/(2N)$ will attain observable frequencies in natural populations. The rate of mutation to such nearly neutral alleles, which will be lower in larger populations, is called the effective neutral mutation rate. The smaller effective mutation rate in larger populations makes the mutational input, $2Nu_e$, less sensitive to N than for strictly neutral mutations. As a consequence, the variation in heterozygosity should be less across species with different population sizes.

Tomoko Ohta is responsible for the theory that most protein evolution is deleterious while most silent evolution is neutral. Her theory was incorporated into what is now called the neutral theory, even though natural selection is playing a role, albeit a negative one. The common element in both theories is that genetic drift is the force responsible for the substitution of alleles. Motoo Kimura once lamented that the theory was not called the "mutation-random drift theory," as this better represents the forces involved in the substitution of alleles.

Derivation of the Fixation Probability

Because much of our understanding of evolution depends on Equation 23, it would be remiss not to discuss its derivation. While the derivation is somewhat more technical than others in this book, it does serve as an entrée into more advanced topics in population genetics.

The derivation of Equation 23 begins with a decomposition of the fixation probability,

$$\pi(p) = \sum_{\Delta p} \text{Prob}\{\Delta p\}\pi(p + \Delta p).$$

(The subscript on π_i will be suppressed in this section, as we will discuss only the fixation probability of the A_1 allele.) Imagine a population with p *as* the initial frequency of the A_1 allele. In the next generation, p will change by a random amount, Δp, whose value reflects the combined action of genetic drift and selection. Suppose you knew with certainty that p changed to a new value p'. Then, for $\pi(p)$ you could use the fixation probability for a population with initial frequency p'. Of course, you will not know what the value of p will be in the next generation. The best you can do is to say that a particular change occurs with some known probability and then average over these changes. This is precisely what the decomposition does. Prob$\{\Delta p\}$ is the probability of a particular change in p, and $\pi(p + \Delta p)$ is the fixation probability for the new frequency. The sum is over all possible changes in p.

The decomposition may be written as

$$\pi(p) = E_{\Delta p}\{\pi(p+\Delta p)\}, \qquad ...(29)$$

which emphasizes that the decomposition really says that the fixation probability in one generation is equal to the expectation of the fixation probability in the next generation (given p in the first generation).

The next step in the derivation uses a Taylor series expansion of the fixation probability in the second generation,

$$\pi(p+\Delta p) \approx \pi(p) + \pi'(p)\Delta p + \frac{1}{2}\pi''(p)(\Delta p)^2,$$

which follows from Equation 1. Plug this into Equation 29 and use the fact that the expectation of a sum is equal to the sum of the expectations to obtain

$$\pi(p) \approx \pi(p) + p'(p)E\{\Delta p\} + \frac{1}{2}\pi''(p)E\{(\Delta p)^2\}.$$

Now subtract $\pi(p)$ from both sides to get

$$\frac{1}{2}\pi''(p)E\{(\Delta p)^2\} + \pi'(p)E(\Delta p\} = 0. \qquad ...(30)$$

The two expectations in Equation 30 are the mean and (approximately) the variance in the change in p given its current value. We studied these moments in the context of pure drift. With selection, the mean change in p *is* not zero, as in the case of pure drift, but rather is approximately

$$E\{\Delta p\} = (s/2)pq = m(p)$$

for the case of additive alleles (m(p) is a commonly used notation for the mean change in p). The expected value of the square of Δp is very nearly equal to the variance in the change in p, because, by

$$\mathrm{E}\{(\Delta p)^2\} = \mathrm{Var}\{\Delta p\} + \mathrm{E}\{\Delta p\}^2,$$

and the square of the mean change in p is very small, by assumption. The variance in Δp due to genetic drift

$$\frac{pq}{2N} = \upsilon(p)$$

($\upsilon(p)$ is commonly used for variance in Δp). Now, Equation 30 may be written as

$$\frac{1}{2}\upsilon(p)\pi''(p) + m(p)\pi'(p) = 0. \qquad ...(31)$$

Equation 31 is a differential equation whose solution, with the proper boundary conditions, is the fixation probability. More exactly, it is a linear second-order differential equation with the two boundary conditions

$$\pi(0) = 0, \ \pi(1) = 1. \qquad \text{...(32)}$$

If the initial value of p is zero, fixation is impossible, hence the first boundary condition. If the initial value is $p = 1$, fixation is a certainty, hence the second.

The solution to Equation 31 subject to boundary conditions 32 is covered in all elementary books on differential equations. Here, we will speed through the solution. Those readers who have trouble with some of the steps should consult a differential equation text.

First, convert Equation 31 to a first-order differential equation by defining

$$f(p) = \pi'(p)$$

(and thus having $f' = \pi''$) and multiply both sides of the equation by $2/\upsilon(p)$ to obtain

$$f'(p) + \frac{2m(p)}{\upsilon(p)} f(p) = 0.$$

This is a first-order linear differential equation and, as such, is much easier to solve than the previous second-order equation. Multiply both sides of the new equation by the integrating factor $\exp(2Nsp)$, and note that the left side is now the derivative of a product ($(u\upsilon)' = u\upsilon' + u'\upsilon$), so

$$\frac{d}{dp}[f(p)e^{2Nsp}] = 0.$$

When the derivative of a function is zero, the function must be a constant, so

$$f(p)e^{2Nsp} = \text{constant}$$

and

$$f(p) = \pi'(p) = c_1 e^{-2Nsp},$$

where c_1 is a constant whose value will be determined when the boundary conditions are imposed.

Finally, integrate the last equation with respect to p to obtain

$$\pi(p) = c_1 \int_0^p e^{-2Nsx} dx + c_2$$

$$= c_1 \frac{1 - e^{-2Nsp}}{2Ns} + c_2.$$

The second constant appears because the indefinite integral is only defined up to an additive constant. (If you differentiate both sides of the equation, you recover the previous one for all values of c_2.) The lower limit of the integral on the right side was chosen for convenience.

To satisfy the boundary condition $\pi(0) = 0$, we require $c_2 = 0$. The boundary condition $\pi(1) = 1$ is satisfied when

$$c_1 = \frac{2Ns}{1 - e^{-2Ns}}.$$

Thus, the fixation probability is

$$\pi(p) = \frac{1 - e^{-2Nsp}}{1 - e^{-2Ns}}$$

as we claimed in Equation 23.

The conceptually important part of the derivation is the original decomposition, which is called a backward equation. A backward equation relates events at a future time (ultimate fixation in our case) to events at the origin of the process. Backward equations are frequently employed in population genetics to learn about the likelihood of different outcomes of evolution as well as learning the time required to reach these outcomes. The reader might enjoy reading Kimura's original derivation of the fixation probability published in 1962.

The derivation brings out the distinction between mean effects in evolution, as captured in the mean function $m(p)$, and variance effects, as captured in $\upsilon(p)$. When setting out to study a new evolutionary model, the first step is usually to describe carefully the mean and variance effects. If the only source of randomness is genetic drift, then $\upsilon(p) = pq/(2N)$ is the proper variance function. However, if the model includes random changes in the environment, then a more complicated variance term must be used. The area of mathematics that gives a description of the evolution of the population once $m(p)$ and $\upsilon(p)$ are known is called diffusion theory. Warren Ewens' short book, *Population Genetics* (1969), is an excellent introduction to the use of diffusion theory in population genetics.

11

QUANTITATIVE GENETICS

Quantitative genetics, an area of scientific enquiry closely allied to population genetics, is concerned with the inheritance of quantitative characters, those whose states, like weight, height, or metabolic rate, fall on a quantitative scale. One of the early triumphs of quantitative genetics was R. A. Fisher's demonstration in 1918 that the correlation between relatives may be explained by the combined action of alleles at many loci, each of relatively small effect. Fisher followed in the footsteps of others who, along with Fisher, put to rest the notion that different laws of inheritance applied to discrete and quantitative traits.

The two traditional concerns of quantitative genetics are the correlation between relatives and the response to selection. Both may be investigated by either a purely statistical approach without explicit reference to the states or frequencies of alleles or from a purely genetic approach that does use allele frequencies, Hardy-Weinberg, and other concepts from population genetics. The former is particularly valuable for introducing the main ideas of quantitative genetics with a minimum of effort and will be developed first. Later, the second approach will be used to explore the effects of dominance on the correlation between relatives.

Our goal is to learn enough of the fundamentals of quantitative genetics to be able to discuss some interesting evolutionary questions that involve quantitative traits.

CORRELATION BETWEEN RELATIVES

Figure is an example of the phenomenology that work in quantitative genetics. In this case, the trait is height and the subjects are students from an evolution class at UC Davis and their parents.

The left side of the figure is a histogram of the heights of the students in the class. As males and females have different average heights, the data have been adjusted to make the two sexes equivalent. Classically, quantitative genetics is concerned with the variation in height rather than with average height. For example, we might ask: How much of the variation in height is due to genes and how much is due to environmental effects? How much is nature and how much nurture? Quantitative genetics is not, in general, concerned with why the average height of the students is approximately 67 inches.

If genes are important, we might expect relatives to resemble each other. More precisely, we might expect a pair of relatives to look more alike than a pair of randomly chosen individuals. The horizontal axis is the midparent value, which is the average height of the two parents. The values are given as deviations from the mean height. A value of –1, for example, implies that the average value of the two parents was 1 inch below the mean height of the population. The vertical axis is the height of the offspring of the two parents that make up the midparent value. Here, too, the heights are deviations from the population mean. There is a clear relationship between midparent and offspring. There is also a lot of scatter around the regression line due, in part, to Mendelian segregation and, in part, to environmental influences. The fact that large parents tend to have large offspring and small parents tend to have small offspring argues that there is a genetic component to height. But how important is that component? The traditional way to pose the question is to ask: Of the total variation in height, what fraction is attributed to genetic causes and what fraction to environmental causes? The answer is the subject of this section.

To assess the relative contribution of genes to a phenotype, we require an explicit model that states, however naively, the way we imagine a phenotype to be constructed from genetic and environmental factors. The simplest such model, in the context of two parents and their offspring, posits a single locus that contributes additively to the phenotype and an environmental component that also acts additively,

$$P = X_m + X_p + \varepsilon. \qquad \text{...(1)}$$

P is the value of the phenotype of an individual expressed as the deviation of its value from the population mean. The symbol X_m refers to the additive effect or contribution of the maternally derived allele to the phenotype. For example, this allele might add 1 inch to the height of its carrier. If so, then both the maternal parent and the offspring will be 1 inch taller because of the additive effects of this allele. X_p refers to the paternally derived allele. Its value will, in general, be different from that of the maternally derived allele. The final component is the contribution of the environment, ε, which is expressed in the same units as the genetic contribution. If we knew, for example, that $X_m = 2$, $X_p = -1$, and $\varepsilon = 5$, then the phenotype would be $P = 6$ units above the population mean.

The mean of P must be zero because all phenotypes are expressed as deviations from the population mean. While this may appear to be a trivial statement, it is worth emphasizing, since we will continually use this fact. The mean values of X_m, X_p, and ε are zero as well. Were they not zero, then the mean value of P would not be zero. Using the notation of expectation, we have

$$\begin{aligned} E\{P\} &= E\{X_m\} + E\{X_p\} + E\{\varepsilon\} \\ &= 0 + 0 + 0 = 0. \end{aligned}$$

(If the expectation operator E is not familiar.)

Of course, we cannot know the actual values of X_m, X_p, and ε for any particular individual. The best we can do is to view these quantities as random variables and estimate their moments in the population. To be precise, we will assume from henceforth that X_m, X_p, and ε are normally distributed random variables with means equal to zero and variances

$$\mathrm{Var}\{X_m\} = \mathrm{Var}\{X_p\} = V_A/2$$

and

$$\mathrm{Var}\{\varepsilon\} = V_E.$$

(The peculiar notation for $\mathrm{Var}\{X_m\}$ and the justification for the normality assumption will be explained shortly.)

The model captured in Equation 1 appears to differ fundamentally from those developed in earlier chapters in that alleles are described by their effects on the phenotype. Earlier, alleles were described as genetic entities with frequencies, but often unspecified phenotypes. In fact, the new model is really the same as the old. Imagine a locus with n alleles at frequencies $p_1, p_2, \ldots, p_n$ that contribute $x_1, x_2, \ldots, x_n$ to the phenotype. The number of alleles is imagined to be so large, and their frequencies so small, that the probability of drawing the same allele twice in a finite sample is small. The mean allelic contribution of this locus is

$$\mu = \sum_{i=1}^{n} x_i p_i,$$

and the variance of the contribution is

$$\sigma^2 = \sum_{i=1}^{n} (x_i - \mu)^2 p_i .$$

As the mean effect of a locus must be zero, $\mu = 0$. The notation for the variance of the the allelic contribution introduced above dictates that $\sigma^2 = V_A/2$. The normality assumption implies that if we were to collect all of the x_i and draw a histogram of their values, that histogram would appear to be a bell-shaped curve. The representation of X in terms of allele frequencies and the x_i emphasizes that X is a population quantity rather than an individual or gene-action quantity. x_i, on the other hand, is the phenotypic contribution of a particular allele and is not a population quantity. A common conceptual error is to regard the additive component as being solely a function of gene action, whereas it is really a function of both gene action as captured in the x_i and the state of the population as captured in the p_i.

The reader may well balk at the suggestion that there are so many alleles at a locus and that their phenotypic contributions are normally distributed. We are making this assumption in order to develop quantitative genetics in the simplest context possible. In the last section of this chapter, we will show how the one-locus model may be replaced with a more realistic multilocus model, where each locus may have only a couple of alleles. The normality then comes, when the number of loci is large, from the Central Limit Theorem.

To take the first steps in the analysis of this model, we need to use variances, covariances, and correlations. If you are not familiar with these moments, this is the time to either read the brief summary or consult a statistics book (or, better yet, to do both). Quantitative genetics is all about variances, covariances, and correlations. If these ideas are not second nature to you, then it is impossible to understand quantitative genetics!

Equation 1 may be used to make precise the question about the relative roles of nature and nurture. As our interest is in explaining variation in quantitative traits, it is quite natural to begin by calculating the variance of both sides of Equation 1,

$$\begin{aligned}\mathrm{Var}\{P\} &= \mathrm{Var}\{X_m\} + \mathrm{Var}\{X_p\} + \mathrm{Var}\{\varepsilon\} \\ &\quad + 2\mathrm{Cov}\{X_m, X_p\} + 2\mathrm{Cov}\{X_m, \varepsilon\} + 2\mathrm{Cov}\{X_p, \varepsilon\}. \qquad ...(2)\end{aligned}$$

The variance in the phenotype is the sum of the variances of the genetic and environmental factors and twice the covariances of these factors. The covariance terms will all be set equal to zero. The term

$\text{Cov}\{X_m, X_p\}$ is the covariance between the maternal and paternal genetic contributions to the phenotype. As the parents are not related (by assumption), their genetic contributions are independent, which implies that their covariance is zero. $\text{Cov}\{X_m, \varepsilon\}$ and $\text{Cov}\{X_p, \varepsilon\}$ are covariances between the genetic and environmental contributions and will be assumed to be zero. Were they not zero, we would say that there are genotype-environment interactions. For example, if one allele added one inch to the phenotype in a warm environment and subtracted one inch in a cold environment and another allele did exactly the opposite, there would be a genotype-environment interaction. Genotype-environment interactions are also called *norms of reaction*. Our only reason for assuming away genotype-environment interactions is to make the model simpler. These interactions do exist in nature but can be minimized in the laboratory.

Without the covariance terms, Equation 2 may be written in the much more pleasing form

$$V_p = V_A + V_E, \qquad \text{...(3)}$$

where

$$V_p = \text{Var}\{P\}$$

is the phenotypic variance,

$$V_A = \text{Var}\{X_m\} + \text{Var}\{X_p\}$$

is the additive variance, and

$$V_E = \text{Var}\{\varepsilon\}$$

is the environmental variance. The additive variance can also be written as

$$V_A = 2\text{Var}\{X_p\} = 2\text{Var}\{X_m\}$$

because the two parental gametes are chosen at random from the population and, for this reason, are statistically equivalent. The "additive" of additive variance refers to the fact that the genetic contribution is a simple sum of the contribution from each allele. In more complicated situations, the two alleles might interact to produce an additional genetic contribution whose variance is called the *dominance variance*.

The fraction of the phenotypic variance due to additive effects is simply

$$h^2 = \frac{V_A}{V_P} = \frac{V_A}{V_A + V_E}, \qquad \text{...(4)}$$

and is called the *heritability* of the trait. For example, if $h^2 = 1/2$, then one-half of the variance in the phenotype is genetic in origin and one-half is environmental. The heritability is precisely the quantity that answers the nature-nurture question. Its value is estimated by using the correlation between relatives, as we will now show.

We begin with an example of a particularly simple pair of relatives: a single parent, mom in this case, and her offspring. The phenotypes of the parent and offspring may be written as follows:

$$P_P = X_m + X'_m + \varepsilon_P$$
$$P_O = X_m + X_p + \varepsilon_O,$$

where the subscripts P and O refer to parent and offspring. The symbol X'_m represents the genetic contribution to the parent's phenotype from the allele that is not passed on to its offspring. With this formulation, the resemblance between parent and offspring is seen to come from the shared allele with phenotypic effect X_m, as X_m is the only common factor on the right sides of both equations. The other alleles in each, X'_m, and X_p, are no more alike in phenotypic effect than are two alleles drawn at random from the population.

The resemblance between parent and offspring is expressed quantitatively by the covariance of their phenotypes, $\text{Cov}\{P_P, P_O\}$, which is

$$\begin{aligned}\text{Cov}\{X_m + X'_m + \varepsilon_P, X_m + X_p + \varepsilon_O\} = \\ \text{Cov}\{X_m, X_m\} + \text{Cov}\{X_m, X_p\} + \text{Cov}\{X'_m, X_m\} + \text{Cov}\{X'_m, X_p\} \\ + \text{Cov}\{X_m, \varepsilon_O\} + \text{Cov}\{X'_m, \varepsilon_O\} + \text{Cov}\{X_m, \varepsilon_P\} + \text{Cov}\{X_p, \varepsilon_P\} \\ + \text{Cov}\{\varepsilon_P, \varepsilon_O\}. \quad \text{...(5)}\end{aligned}$$

The fact that the covariance of a sum is the sum of the covariances. Although the covariance looks horrendous, it is important to appreciate that conceptually it is quite simple. The covariance between parent and offspring must ultimately be due to the covariances of the additive components of each phenotype. There must necessarily be a lot of covariance terms because there are three random quantities contributing to each phenotype. Fortunately, most of these covariances are zero, either by assumption or because of independence, as we will now show.

The first line of covariances on the right side of Equation 5 are between the genetic contributions to the phenotypes. The first term in this line is the covariance of a random variable with itself, which is its variance,

$$\text{Cov}\{X_m, X_m\} = \text{Var}\{X_m\} = V_A/2.$$

The next three covariances are between independently derived alleles. As these alleles are unrelated, the covariances of their effects are zero. The next line has covariances between a genetic effect and an environmental effect; they are zero because of our previous assumption of no genotype-environment interactions.

Finally, we come to the thorny Cov$\{\varepsilon_P, \varepsilon_O\}$, which is the covariance of the environmental effects of parent and offspring. This, too, will be set equal to zero, but not without some misgivings. Often, the environments of parents and offspring are more similar than are those between randomly chosen individuals. This is certainly true for humans, where geography, socio-economic class, level of education, and a myriad of other causes of common environments occur. Common environments will cause relatives to resemble each other more than they would otherwise. As a consequence, common environments may give an experimenter an inflated view of the role of genetics in the determination of the trait. In laboratory situations, the common environment can be minimized and, as most quantitative genetics experiments are done in the laboratory, will be set to zero here.

The expression for the covariance of parent and offspring is now the remarkably simple

$$\text{Cov}\{P_P, P_O\} = \frac{V_A}{2}. \qquad \text{...(6)}$$

The covariance between parent and offspring is one-half the additive variance. The presence of the additive variance is traceable to the fact that both the parent and offspring share the allele with value X_m, whose variance is $V_A/2$.

The correlation coefficient between a parent and its offspring is obtained by dividing both sides of Equation 6 by V_P and using the definition of the correlation coefficient

$$\begin{aligned}\text{Corr}\{P_P, P_O\} &= \frac{\text{Cov}\{P_P, P_O\}}{\sqrt{\text{Var}\{P_P\}\text{Var}\{P_O\}}} \\ &= \frac{V_A}{2V_P} = \frac{h^2}{2}.\end{aligned}$$

(In the second line we used the fact that the phenotypic variances of any set of unrelated individuals are the same. Thus, the phenotypic variance of parents is the same as that of offspring.) The correlation is an increasing function of the additive variance. For a fixed V_E, the larger the additive variance the greater the resemblance between relatives. Without genetic diversity, relatives would look no more alike than random pairs of individuals. With genetic diversity, they resemble each other because their shared alleles are more similar than random pairs of alleles. The next task is to find the correlation between an arbitrary pair of relatives, X and Y, whose phenotypes are determined by the two equations

$$P_X = X_m + X_p + \varepsilon_X$$
$$P_Y = Y_m + Y_p + \varepsilon_Y.$$

If we charge ahead, ignoring all genotype-environment interaction and common environment terms, we obtain

$$\text{Cov}\{P_X, P_Y\} = \text{Cov}\{X_m, Y_m\} + \text{Cov}\{X_m, Y_p\} + \text{Cov}\{X_p, Y_m\} + \text{Cov}\{X_p, Y_p\}. \quad \text{...(7)}$$

The values of the covariances depend on the average number of shared alleles between the relatives. With probability r_0, the two relatives share no identical-by-descent alleles and all of the covariances in Equation 7 are zero. With probability r_1, the two relatives share one pair of identical-by-descent alleles, in which case only the covariance corresponding to the shared allele will be non-zero, and its value is $V_A/2$. (Note that this case is like parent and offspring, for which $r_1 = 1$.) Finally, the relatives could share two identical-by-descent alleles. In this case each pair of identical-by-descent alleles yields one non-zero covariance of magnitude $V_A/2$, so the full covariance when there are two pairs of identical-by-descent alleles is V_A. Considering all three contingencies, the covariance between relatives X and Y is

$$\text{Cov}(P_X, P_Y) = r_0 \times 0 + r_1 \times \frac{V_A}{2} + r_2 \times V_A$$

or, using the coefficient of relatedness, $r = (r_1/2) + r_2$,

$$\text{Cov}(P_X, P_Y) = rV_A. \quad \text{...(8)}$$

The correlation between the relatives is obtained by dividing both sides of Equation 8 by the phenotypic variance,

$$\text{Corr}\{P_X, P_Y\} = rh^2 \quad \text{...(9)}$$

In words, the correlation between a pair of relatives is equal to the coefficient of relatedness times the heritability. This is certainly one of the most satisfying of answers to an ostensibly difficult problem in all of biology. In his 1918 paper, R.A. Fisher derived the correlation between relatives as a series of special cases for many representative pairs of relatives. The general form of the correlation, of which Equation 9 is a special case, was derived in the 1940s by Gustave Malecot and Charles Cotterman.

Heritabilities are easy things to measure. As a consequence, the literature is full of heritability estimates for almost any trait you can imagine. Two important generalities emerge from heritability studies. The first is that almost all traits have heritabilities between 0.1 and 0.9. In other words, between 10 and 90 percent of the phenotypic variation seen in most quantitative traits is genetic in origin, an observation that echoes the ubiquitous variation seen in DNA and proteins and poses yet again the question of why so much genetic variation exists.

Table 11.1. Heritability estimates determined by parent-offspring correlations for a variety of traits and species.

Species	*Character*	*Heritability*
Honeybee	oxygen consumption	0.15
Eurytemora herdmani	length	0.12
Cricket	wing length	0.74
Flour beetle	fecundity	0.36
Red-backed salamander	vertebral count	0.61
Darwin's finch	weight	0.91
Darwin's finch	bill length	0.85

The second generalization is that life history traits, like viability, longevity, and fecundity, tend to have lower heritabilities (average $h^2 = 0.12$) than do behavioural traits (0.17), which tend to have lower heritabilities than morphological traits (0.32). The reason for this pattern is obscure. Life history traits could have lower heritabilities because of lower additive variances or because of higher environmental variances. The folklore of our field claims that life history traits have lower additive variances because they are closely tied to fitness and that natural selection quickly and efficiently removes additive variation in fitness. However, Dave Houle (1992) recently showed that life history traits actually have higher additive variances than other traits. Apparently, life history traits hive lower heritabilities because the joint contributions of environmental, dominance, and epistatic variances are much larger than those in other traits. So much for folklore!

Closely related to our discussion of the correlation between relatives is the answer to the following question: If the phenotype of relative X is $P_X = x$, then what is the expected phenotype of relative Y? For example, if I knew that the average height of a student's parents was 72 inches, what would be the expected height of the student? The answer comes immediately from the regression coefficient. The regression coefficient is the slope of the "best fit". The line, which is forced to pass through the origin, is fit to the data by the method of least squares, which minimizes the squared deviation (measured vertically) of each point from the line. The slope of the regression line is

$$\beta = \frac{\text{Cov}\{P_X, P_Y\}}{\text{Var}\{P_X\}} = rh^2,$$

using the definition of the regression coefficient in Equation 8. Thus, the regression of one relative on another is the same as the correlation coefficient of the relatives.

The expected value of relative Y, given that the phenotype of relative X is x, is

$$\text{E}\{\text{P}_{\text{Y}} | P_X = x\} = \beta x = rh^2 x,$$

where, as usual, all measurements are expressed as deviations from the mean. Note that the above relationship holds only if the trait is normally distributed.

The most important implication of this result is that the Y relative is always closer to the population mean, on average, than is the X relative because both r and h^2 are less than one (in most situations). The movement of the mean of relative Y toward the population mean is called the regression toward the mean. It occurs because a pair of relatives share only a portion of their alleles. The rest are obtained at random from the population. Thus, the mean of one relative, given the value of the other, must move toward the population mean.

What is the heritability of height? One of the "relatives" in the figure is a midparent, which is the average of the phenotypes of the two parents of each student. Consequently, we cannot use the results developed thus far without some minor modifications. The covariance of midparent and offspring is

$$\begin{aligned}\text{Cov}\{(P_P + P_{P'})/2, P_O\} &= \text{Cov}\{P_{P'}, P_O\}/2 + \text{Cov}\{P_{P'}, P_O\}/2 \\ &= V_A/2,\end{aligned}$$

just as it is for one parent and offspring. However, the variance of midparent is smaller than that of a single parent,

$$\text{Var}\{(P_P + P_{P'})/2\} = V_P/2.$$

The correlation of midparent and offspring is

$$\frac{V_A/2}{\sqrt{V_P V_P/2}} = \frac{1}{\sqrt{2}}\frac{V_A}{V_P} = \frac{1}{\sqrt{2}}h^2.$$

Similarly, the regression of offspring on midparent is

$$\frac{\text{Cov}\{(P_P + V_{P'})/2, P_O\}}{\text{Var}\{(P_P + P_{P'})/2\}} = \frac{V_A/2}{V_P/2} = h^2, \qquad \text{...(10)}$$

which is not the same as the correlation coefficient for normal pairs of relatives. Table 11.2 is a handy summary of the various measures of relationship between midparent and offspring and those of other pairs of relatives.

Table 11.2. A summary of the measures of resemblance between pairs of relatives.

	Parent-offspring	*Midparent-offspring*	*General*
Covariance	$V_A/2$	$V_A/2$	rV_A
Correlation	$h^2/2$	$h^2/\sqrt{2}$	rh^2
Regression coefficient	$h^2/2$	h^2	rh^2

Response to Selection

Agriculturalists, from prehistoric times until the present, have improved their crops and livestock by selective breeding: simply choose the best individuals as parents of the next generation. And it works, providing that there is additive variance for the traits of interest. But how well does selective breeding work? Will there be a significant change in the mean value of a trait in a few generations, or are hundreds or even thousands of generations of selective breeding required? The answer is the main goal of this section.

Natural selection on quantitative traits is like selective breeding in that more adaptive morphologies are more likely to leave behind offspring than are less adaptive morphologies. Here, we will see how fitness differences between morphologies change the distribution of morphologies. Surprisingly, we can do this, at least approximately, without explicitly describing changes in genotype frequencies.

In this case the character is the number of bristles on the fourth and fifth abdominal sternites in *Drosophila melanogaster*. Bristle number in *Drosophila* has been studied for years simply because it is easier to count accurately the number of bristles under a microscope than it is to determine more universal quantitative traits like weight or length. The graph shows the mean number of bristles in lines selected for high or low numbers of bristles for five generations. Five replicates lines, (the upper five), were selected for high bristle number and five (the lower five) were selected for low bristle numbers. The selection was performed by counting bristles on 100 males and 100 females in each line and then choosing for parents the 20 males and 20 females with the highest and lowest numbers of bristles for their sex. After five generations, selection was stopped and the bristles were not counted again until generation 24.

The selection experiment was obviously successful in that the number of bristles increased steadily in the high lines and decreased steadily in the low lines. After five generations, a typical fly in the high lines had about 13 more bristles than the original population, while the low lines had about 7 fewer bristles. This asymmetry in the response to equal selection in both directions is seen in almost all selection experiments.

After five generations, selection was relaxed and the number of bristles moved back toward the number in the original population. This, too, is typical of other selection experiments. One possible explanation is that there is an optimal number of bristles and that natural selection moves the population back toward the optimum once artificial selection is stopped. Another possibility is that the alleles whose frequencies increased in the population as a result of selective breeding have pleiotropic deleterious effects or are linked to deleterious alleles. Once artificial selection stops, natural selection acts on these pleiotropic, or linked, factors.

Our first task is to find a quantitative description of the expected progress of selection in a typical selective breeding experiment. The starting point, the top curve, is a population of potential parents whose phenotypes are, by assumption, normally distributed. From this population, a group of individuals are chosen to serve as parents for the next generation. The usual practice is to pick a fixed number of individuals to measure and then choose a fixed proportion of these as parents. For example, in the *Drosophila* experiment 100 individuals of each sex were measured and the top and bottom 20 percent were chosen as parents. The chosen parents are then put into male-female pairs and the midparent value of each pair is noted. The mean midparent value of the selected parents, expressed (as always) as the deviation from the population mean, is called the selection differential and is notated by S.

The distribution of the offspring of the selected parents is illustrated in the next line of the figure. As the mean of the selected parents is larger than that of the potential parents, the distribution of the offspring is shifted to the right. The deviation of the offspring mean from the potential parent mean is called the selection response and is notated by R.

What is the relationship between the selection differential, S, and the selection response, R? We argued that the regression of offspring on midparent is the heritability. It follows that the mean value of the offspring of a mating with midparent value $P_{MP} = x$ is h^2x. In the selection experiment, there are several pairs of selected parents, each with its own midparent value, say x_i for the *i*th pair of parents. The expected phenotype of the *i*th pair of parents is thus h^2x_i. In a selection experiment with *n* pairs of parents, excepted phenotype of the offspring is

$$\frac{1}{n}\sum_i E\{P_O | P_{MP} = x_i\} = h^2 \frac{1}{n}\sum_i x_i.$$

The term on the left side is the average expected value of the offspring (expressed as a deviation from the parental mean), which is the selection response, *R*. The sum on the right side is the average value of the phenotypes of the selected parents, or the selection differential, *S*. In other words, the selection response is simply

$$R = h^2S \qquad ...(11)$$

If the heritability of a trait is one-half, then the response of one generation of selection is to move the population mean halfway between its value in the parental generation and the mean value of the selected parents.

Equation 11 is the answer to the question posed at the beginning of this section: How well does selective breeding work? Like the equation for the cor-relation between relatives, it is a remarkably simple answer to an ostensibly difficult question. However, while $R = h^2S$ is an accurate description of the response to selection in a single generation, it is not necessarily an accurate predictor of the progress of selection over several successive generations because each generation of selection changes h^2 in ways that are impossible to predict. We will return to this problem after discovering another way to write the selec-tion differential.

When designing a selection experiment, it is often useful to know the propor-tion of parents needed to obtain a particular selection differential. In fact, the *Drosophila* bristle-number experiment described at the beginning of this sec-tion was couched entirely in terms of the proportion selected. We cannot relate that experiment to Equation 11 without first finding a connection between the proportion selection and the selection differential. The proportion selected is the area of the shaded portion of the distribution of parental pheno-types. The mean value of the shaded portion is the selection differential. The connection between the two is given by the formula

$$S = \sqrt{V_P}\, i(p), \qquad ...(12)$$

where the function $i(p)$ is called the intensity of selection. Equations 11 and 12 together give a new equation for the response to selection,

$$R = h^2 i(p)\sqrt{V_P}\,, \qquad \text{...(13)}$$

which is particularly convenient for many applications.

The 1957 paper by George Clayton, J. A. Morris, and Alan Robertson, also provided one of the first demonstrations that Equation 13 can accurately describe the response to selection in favourable experimental settings. Clayton et al. did this by estimating the heritability and phenotypic variance of bristle number and then checking how well Equation 13 predicts the progress of selection for different fractions of selected individuals. The heritability for bristles is about $h^2 = 0.52$, and the phenotypic variance, an average of the male and female variances, is $V_P = 11.223$, so $\sqrt{V_P} = 3.35$. For the first line in the table, where the intensity of selection is 1.40, the predicted response to selection is

$$R = 0.52 \times 1.40 \times 3.35 = 2.42,$$

which is close to the observed change upward of 2.62 bristles per generation. The response to selection for other values of the intensity of selection in the high lines is also in rough agreement with the theoretical predictions. But what about the low lines? Note that there is nothing in our theory that treats up and down selection differently. Thus, the down line should decrease by about 2.42 bristles per generation for the highest intensity of selection. In fact, it decreased only by 1.48 bristles. The authors of the paper are at a loss to explain this discrepancy. They do point out that the observed change per generation is obtained by averaging the changes for each of the five generations. In the case of the low lines, they argue that the response to selection declined after the first couple of generations of selection and, by implication, that h^2 declined. However, there is little support for this explanation, where the low lines appear to decrease linearly for all five generations of selection. Other possible explanations include the possibility that natural selection opposes artificial selection for lower bristle numbers or that there is a scaling effect due to an asymmetry in the distribution of additive effects on bristle number.

The suggestion that heritabilities decrease during selection points out one of the major obstacles to using Equation 13 to predict the response to selection for more than a few generations into the future. Each generation of selection changes the genetic structure of the population by changing allele frequencies and the associations of alleles

Table 11.3. The predicted and average observed change in the number of bristles in *Drosophila* in a single generation for different intensities of selection.

p	*i(p)*	*Predicted*	*Observed up*	*Observed down*
20/100	1.40	2.42	2.62	1.48
20/75	1.24	2.14	2.20	1.26
20/50	0.97	1.68	1.46	0.79
20/25	0.35	0.61	0.28	–0.08

on chromosomes. As a consequence, the additive variance and the heritability are likely to change and, with them, the response to selection. Unless there is some reason to believe that heritabilities remain relatively constant, perhaps because selection is very weak and mutation is restoring any lost variation, one should not use Equation 13 to predict events after the first few generations of selection.

The brevity of this section shows how simply the response to selection follows from the regression of offspring on midparent. However, it belies the considerable complications that arise in actual selection experiments. For a thorough treatment of the response to selection, there is no better place to turn than Douglas Falconer's *Introduction to Quantitative Genetics* (1989).

Evolutionary Quantitative Genetics

Quantitative traits pose the same sorts of evolutionary questions as do discrete traits. Just as we puzzled over ubiquitous molecular variation—our Great Obsession—we should wonder why there is so much genetically determined variation in quantitative traits. Similarly, arguments over the strength of selection acting on molecular variation have their counterparts in arguments over the strength of selection acting on quantitative traits. In this section, we will just brush the surface of these questions, which are not nearly as well posed for quantitative traits as they are for molecular variation because we do not understand the genetic basis of quantitative variation. This difference is offset somewhat by the fact that quantitative traits are under much stronger selection than are molecular traits, making it easier to guess the adaptive value of many quantitative traits.

Quantitative genetics has been used to reconstruct the selective forces acting on quantitative traits. The basic idea is to turn our previous discussion on its head by using the selection response to estimate the selection differential. An interesting example is in a paper by Russ Lande (1976), which examines selection on paracone height in fossil horses. In the course of their radiation from the Eocene to the Pleistocene, horses moved from eating leaves in forests to eating grasses in plains. As eating grasses wears down teeth faster than eating leaves, horses evolved higher ridges, called paracones, on their teeth. The change in paracone height over an interval of time of length T is called R_T. The ratio of R_T to the phenotypic standard deviation, $\sqrt{V_P}$, for pairs of species is given in the second column. The time of separation of the pairs is given in the third column.

Table 11.4. The fraction of selective deaths, *p*, required to account for the evolution of paracone height in horses under the assumption of directional selection.

From species to species	$R_T/\sqrt{V_P}$	T	$1-p$
Hyracotherium to *Mesohippus*	10.6	10×10^6	4×10^{-7}
Mesohippus to *Merychippus*	25.6	5×10^6	2×10^{-6}
Merychippus to *Neohippus*	7.8	1.75×10^6	2×10^{-6}
Hyracotherium to *Neohippus*	44.0	16.75×10^6	1×10^{-6}

Let us assume that the increase in paracone height was due to a constant directional selection pressure and try to find the strength of selection. The selection response may be calculated by first rearranging Equation 13 to

$$\frac{R}{\sqrt{V_P}} = h^2 i(p).$$

The response per generation, assuming that a horse generation is one year, is obtained by dividing the total difference in paracone height by the length of. separation, T, $R = R_T/T$. The heritability of paracone height is assumed to be one-half, a typical figure for morphologic traits. Thus, the intensity of selection becomes

$$i(p) = \frac{2R_T}{T\sqrt{V_P}}.$$

The parameters on the right side may be obtained from Table 5.4 to arrive at the intensity of selection. For example, from the first line the intensity of selection is

$$i(p) = \frac{2 \times 10.6}{10^7} = 2.12 \times 10^{-6}.$$

The value of p corresponding to the intensity of selection could, in principle, be obtained. However, the resolution of the graph for such a small intensity of selection is not sufficient. The values of 1 – *p*, obtained by Lande using approximations to the integrals appearing in the definition of the intensity of selection.

The values of 1 – *p*, which are the proportion of horses not included as parents each generation, suggest that only about one to four horses of every million need suffer a selective death in order to evolve paracone height at the rate seen in the fossil record. This is incredibly weak selection! So weak, in fact, that genetic drift could well dominate selection. Does this mean that we should entertain drift *as* a candidate to explain the evolution of paracone height? Perhaps, but a much more profitable direction is to accept that our assumption of constant directional selection is unrealistic. Paracone height is probably under stabilizing selection for an optimum height that is determined by, among other factors, the time spent feeding on grasses and the nature of the grasses. The optimum itself will slowly change upward as the grasslands begin to dominant the landscape. Instead of a picture of an excruciatingly slow response to a distant selective goal, we picture very strong selection keeping the species right at the optimum paracone height. Under this hypothesis, the slow increase in paracone height is due to the slow change in the optimum rather than to a slow response to very weak selection.

The investigation of paracone height poses the questions: How fast have quantitative traits evolved in the fossil record? How do these rates of evolution compare to rates observed under artificial selection? Is the rate of evolution of paracone height typical or unusually fast or slow? J.B.S. Haldane, one of the pioneers of population genetics, proposed that rates of phenotypic evolution be quantified by

$$k_h = \log(x_2/x_1)/T \text{ Darwins.}$$

In this expression, x_1 and x_2 are measures of some trait for two species on the same lineage that are separated by T million years. The units of k_h are called Darwins. For example, the paracone height of *Hyracotherium* is $x_1 = 1.54$ and of *Mesohippus* is $x_2 = 2.12$. As these two horses are separated by 20 million years, the rate of paracone evolution is

$$k_h = \frac{\log(2.12 / 1.54)}{20} = 0.016 \text{ Darwins.}$$

The most striking feature of the table is that the rate of evolution in selection experiments is at least two orders of magnitude faster than any rate seen in nature or the fossil record. The fastest rates in nature are for recent colonizers, which might be expected to evolve

Table 11.5. Rates of morphological evolution. The time intervals are measured in either years (yr) or millions of years (MY).

	Number	*Time interval*	*Rate (Darwins)*
Selection experiments	8	3.7 yr	58.700
Recent colonizations	104	170 yr	370
Post-Pleistocene mammals	46	8200 yr	3.7
Fossil vertebrates	228	1.6 MY	0.08
Fossil invertebrates	135	7.9 MY	0.07

quickly to adapt to their new environments. Paracone height is clearly evolving very slowly compared to what is possible in the laboratory or even what is seen as the fastest rates in nature. This lends support to the view that the evolution of paracone height is due to a slowly changing optimum rather than directional evolution to a radical new environment. Thus, when considering the applicability of selection theory to the fossil record, one should never assume that the selective forces have remained constant for appreciable periods of time.

The Lande paper is a quantitative genetics analogue to the study of rates of molecular evolution. Similarly, there is a quantitative genetics version of the Great Obsession: Why is there so much genetic variation for quantitative traits? The study of quantitative genetic variation is very similar to that of molecular variation. As a first step, for example, one might try to find the amount of quantitative variation introduced each generation by mutation and the amount removed by genetic drift to see if the equilibrium determined by these two forces is compatible with observed levels of quantitative variation, just as Kimura and Ohta did for molecular variation.

A seminal paper on quantitative genetic variation is a short work by George Clayton and Alan Robertson published in 1955. Their main interest was the extent to which mutations may contribute to the overall response to selection in experiments like that of Clayton, Morris, and Robertson described in the previous section. Before their paper, the relative roles of allele frequency change and mutational input to the selection response were not well understood theoretically or empirically. We begin by first describing the experiment and its analysis, and then we will discuss its implications for the question of the maintenance of variation.

The experiment used a straightforward selective breeding protocol. The only novel feature was the initial population, which was nearly homozygous because of many generations of brother-sister mating. With so little initial variation, the response to selection should be due, in large part, to newly arising mutations affecting bristle number. High and low lines were selected by choosing the 10 most extreme individuals out of 25 for each sex, which gives $i(0.4) = 0.94$ for the intensity of selection. Such selection continued for 14 generations. By averaging the high and low lines across all 14 generations, Clayton and Robertson concluded that the average response in one generation of selection was $R = 0.027$ bristles. Using Equation 13 and their estimate that $V_P =$ 4.41, the average additive variance over the 14 generations is

$$\bar{V}_A = 0.06,$$

where the bar over V_A is a reminder that this is an average over generations.

In the course of the experiment, genetic drift removes additive variance at a rate of $1/(2N)$, and mutation augments the additive variance by an amount V_m each generation. V_m is the quantity that we want to estimate. Thus, the model for the evolution of V_A during the 14 generations is,

$$V_A(t) = \left(1 - \frac{1}{2N}\right) V_A(t-1) + V_m, \qquad ...(14)$$

which appears in the text of the paper, albeit with different notation. We will not derive this equation, although it should be in accord with your intuition. In each generation, genetic drift reduces the additive variance by the factor $1-1/(2N)$, just as heterozygosity is reduced by this factor when discussing molecular variation. In each generation, mutation augments the additive variance by V_m, which is analogous to the mutational input, $2Nu$, of molecular variation.

Equation 14 can be solved iteratively, once we know the 'additive variance that was present at the start of the experiment, $V_A(0)$. The authors chose not to estimate $V_A(0)$ directly, but rather to argue that additive variance is at an equilibrium between its removal by inbreeding (brother-sister mating) and its introduction by mutation. Brother-sister mating removes variance at the rate 0.191. Thus, at equilibrium

$$0.191 V_A(0) = V_m,$$

or

$$V_A(0) = 5.2 V_m.$$

Equation 14 may now be iterated to find the additive variance in successive generations. For generation 1,

$$V_A(1) = 0.975 \times 5.2 V_m + V_m = 6.07 V_m,$$

using a population size of $N = 20$, for which

$$1-\frac{1}{2N}=0.975.$$

The next generation is

$$V_A(2) = 0.975 \times 6.07V_m + V_m = 6.92V_m.$$

Continuing in this way, $V_A(14) = 15V_m$. The average additive variance (add up the $V_A(i)$ and divide by 14) is

$$V_A = 10V_m = 0.06,$$

from which we conclude that $V_m = 0.006$. In a typical *Drosophila* population, $V_A = 5$. Thus, mutation augments the additive variance by about 0.006/5, or 0.1 percent its value each generation.

Clearly, the mutational input to this particular quantitative trait each generation is very small. To account for the observed value of V_A in natural populations, we must invoke a very weak force to remove the variation. Genetic drift is always a good candidate for a weak force, providing that we are willing to accept that bristle number is a neutral trait. If so, we can use Equation 14 to argue that the equilibrium additive variance will satisfy

$$\hat{V}_A=\left(1-\frac{1}{2N}\right)\hat{V}_A+V_m,$$

which gives

$$\hat{V}_A=2NV_m.$$

Recalling that $V_m/V_A = 0.001$, we must conclude that the effective size of *Drosophila* is 500, an absurdity. The additive variance in natural populations is far less than that predicted by this model.

Where did we go wrong? Many evolutionists would argue that bristle number is not a neutral trait. Rather, there is an optimum number of bristles, and mutations that move a fly away from the optimum are selected against. This would have the effect of lowering the additive variance well below that predicted by the neutral models. Others would argue differently. They would say that bristle number really is a neutral trait, but that mutations changing the numbers of bristles affect other traits as well, and these other traits are under selection. In this case, there is selection against bristle mutations because of their pleiotropic effects. Currently, we do not know which of these two explanations is closer to reality; perhaps neither is accurate, and some other set of forces is maintaining additive variance.

Dominance

This final section introduces an approach to quantitative genetics based on only two alleles at a locus rather than the large number of alleles used. There are two immediate benefits. The first is that dominance may be included in away that more nearly fits our biological sense of dominance than would be possible with the statistical approach of the first section. The second is that we will be able to show that the genetic contribution to traits affected by many di-allelic loci, each of small effect, is approximately a normal distribution. Thus, this section could be viewed as a genetically based justification for the approach taken in the first section.

With dominance, the linear model for the construction of a phenotype is

$$P = X_m + X_p + X_{mp} + \varepsilon,$$

where the new term, X_{mp}, captures the dominance relationships between the maternally and paternally derived alleles. If V_D is the variance in X_{mp}, then, assuming that the additive and dominance contributions are uncorrelated,

$$V_P = V_A + V_D + V_E.$$

There is no obvious reason why the additive and dominance contributions should not be correlated. In fact, a little reflection shows that we haven't said precisely what is meant by the additive and dominance contributions. The confusion will be cleared up in this section. We will show that the additive effects are chosen in such a way as to maximize their contributions to the phenotype while assuring that the additive effects are uncorrelated with the dominance effects.

To start, we need to define the contributions to the phenotype made by a single locus with two alleles:

Genotype:	A_1A_1	A_1A_2	A_2A_2
Frequency:	p^2	$2pq$	q^2
Genotypic value:	a_{11}	a_{12}	a_{22}
Additive value:	$2\alpha_1$	$\alpha_1 + \alpha_2$	$2\alpha_2$

The genotypic values are the contributions of the genotypes to the phenotype expressed as deviations from the population mean. Consequently, the mean of the genotypic values must be zero,

$$p^2a_{11} + 2pqa_{12} + q^2a_{22} = 0.$$

The additive values of the three genotypes are values that come as close as possible to making this locus one with no dominance. The additive values depend of the genotypic values and the allele frequencies. They will be derived shortly. For now, we will note only that the additive values are expressed as deviations from the population mean, so the mean of the additive effects must be zero as well,

$$p^2 2\alpha_1 + 2pq(\alpha_1 + \alpha_2) + q^2 2\alpha_2 = 2\alpha_1 p + 2\alpha_2 q = 0. \quad \text{...(15)}$$

The variance of the genetic contribution to the phenotype, the genetic variance, is

$$V_G = p^2 a_{11}^2 + 2pq a_{12}^2 + q^2 a_{22}^2, \quad \text{...(16)}$$

recalling that the mean the genotypic values is zero. The additive variance is given by

$$\begin{aligned} V_A &= p^2 (2\alpha_1)^2 + 2pq(\alpha_1 + \alpha_2)^2 + q^2 (2\alpha_2)^2 \\ &= 2p^2 \alpha_1^2 + 2(p\alpha_1 + q\alpha_2)^2 + 2q\alpha_2^2 \\ &= 2(p\alpha_1^2 + q\alpha_2^2), \end{aligned} \quad \text{...(17)}$$

where the derivation of the final form requires Equation 15. At this point, however, we do not know the additive values (α_i, is function of a_{ij} and *p*); we will return to V_A once we do.

The horizontal axis is the number of A_1 alleles in the genotype. With each new A_1 allele, the additive value of the genotype is increased by α_1, as illustrated by the sloping line. The deviation of the genotypic values from the additive values is illustrated by the vertical lines connecting the two. The equation for the line is obtained by minimizing the mean squared deviation of the line from genotypic values,

$$Q(\alpha_1, \alpha_2) = p^2 (a_{11} - 2\alpha_1)^2 + 2pq(a_{12} - \alpha_1 - \alpha_2)^2 + q^2 (a_{22} - 2\alpha_2)^2,$$

coupled with the previous assumption that the mean of the additive effects is zero. Finding a best-fit line by minimizing the squared deviation is called the method of least squares. The advantage of this particular method is that the additive and dominance effects are uncorrelated, as you will show in a problem once you have all of the requisite formulae.

The additive effects are found by minimizing $Q(\alpha_1, \alpha_2)$ using the usual method of setting the derivatives of Q equal to zero,

$$\frac{\partial Q}{\partial \alpha_1} = -4[p^2 (a_{11} - 2\alpha_1) + pq(a_{12} - {}_1 - \alpha_2)] = 0$$

$$\frac{\partial Q}{\partial \alpha_2} = -4[pq(a_{12} - \alpha_1 - \alpha_2) + q^2 (a_{22} - 2\alpha_2)] = 0.$$

Both equations are readily simplified by first isolating the mean additive effect, $p\alpha_1 + q\alpha_2$, and setting it equal to zero and then dividing the first equation by $-4p$ and the second equation by $-4q$. The resulting equations may be solved to obtain

$$\alpha_1 = pa_{11} + qa_{12} \quad \text{...(18)}$$

$$\alpha_2 = pa_{12} + qa_{22}. \qquad ...(19)$$

The additive values may now be substituted into Equation 17 to obtain the additive variance in terms of the genotypic values,

$$V_A = 2pq[p(a_{11} - a_{12}) + q(a_{12} - a_{22})]^2. \qquad ...(20)$$

Finally, the dominance variance is obtained by subtracting the additive variance from the genetic variance,

$$V_D = V_G - V_A = p^2q^2(a_{11} - 2a_{12} + a_{22})^2. \qquad ...(21)$$

Thus, we have completed our partitioning of the total genetic variance into additive plus dominance components,

$$V_G = V_A + V_D. \qquad ...(22)$$

The next step is to find the genetic contribution to the covariance between an arbitrary pair of relatives. The approach is mundane, but the algebra is cumbersome. Recall that the covariance between two random variables is

$$\sum_i \sum_j p_{ij} x_i y_j.$$

In our case, the random variables are the genotypic values of relatives X and Y as summarized here. For example, the first line gives the pertinent data for a pair of relatives, both of whom are A_1A_1. The contribution of this locus to the phenotype of both relatives is a_{11}. The frequency of this pair is the probability that X is A_1A_1, p^2, times the probability that Y is A_1A_1 given that X is A_1A_1. The latter probability is sum of the probabilities of three mutually exclusive events. The first event is when X and Y share no identical-by-descent alleles, r_0, times the probability that Y is A_1A_1 given that it shares no identical-by-descent alleles with X, p^2. The probabilities of the other two events are obtained in a similar fashion, as are the rest of the frequencies in the table.

The covariance is calculated by adding up nine terms, each of which is the frequency of a pair times the product of the two values of that pair. The sum is simplified by considering, in turn, the coefficients of the r_i. The coefficient of r_o is

$$(p^2a_{11} + 2pqa_{12} + q^2a_{22})^2,$$

which is just the square of the average genotypic value, which is zero.

The coefficient r_1, after a modest bit of algebra that requires the use of equation 18, is

$$p\alpha_1^2 + q\alpha_2^2,$$

Table 11.6. All possible pairs of relatives needed for the calculation of the covariance between relatives.

Genotypes		*Values*		
X	*Y*	*X*	*Y*	*Frequency*
A_1A_1	A_1A_1	a_{11}	a_{11}	$p^2[r_0p^2 + r_1p + r_2]$
A_1A_1	A_1A_2	a_{11}	a_{12}	$p^2[r_0pq + r_1q]$
A_1A_1	A_2A_2	a_{11}	a_{22}	$p^2[r_0p^2]$
A_1A_2	A_1A_1	a_{12}	a_{11}	$2pq[r_0p^2 + r_1p/2]$
A_1A_2	A_1A_2	a_{12}	a_{12}	$2pq[r_02pq + r_1/2 + r_2]$
A_1A_2	A_2A_2	a_{12}	a_{22}	$2pq[r_0q^2 + r_1q/2]$
A_2A_2	A_1A_1	a_{22}	a_{11}	$q^2[r_0p^2]$
A_2A_2	A_1A_2	a_{22}	a_{12}	$q^2[r_0pq + r_1p]$
A_2A_2	A_2A_2	a_{22}	a_{22}	$q^2[r_0q^2 + r_1q + r_2]$

which, by Equation 17, is $V_A/2$. Finally, the coefficient of r_2 is, from Equation 16, V_G. Taken together, these calculations show that

$$\text{Cov}\{P_X, P_Y\} = (r_1/2)V_A + r_2V_G,$$

or

$$\text{Cov}\{P_X, P_Y\} = (r_1/2 + r_2)V_A + r_2V_D$$
$$= rV_A + r_2\ V_D. \qquad \text{...(23)}$$

When there is no dominance, this is the same as Equation 8, which was obtained by a different method. In fact, the two derivations share one critical feature: they both make explicit use of the resemblance that comes from sharing identical-by-descent alleles.

Previously, we assumed that there was no dominance and used the model

$$P = X_m + X_p + \varepsilon$$

to partition variances and find the correlation between relatives. With dominance, the model was extended to

$$P = X_m + X_p + X_{mp} + \varepsilon, \qquad \text{...(24)}$$

where the new term, X_{mp}, represents the contribution of dominance to the phenotype. As with X_m and X_p, X_{mp} is a population quantity that depends on allele frequencies. With just one locus and two alleles, it is a simple matter to describe the universe of genetic effects. Furthermore, we can partition the phenotypic variance into

$$V_P = V_A + V_D + V_E,$$

providing that we assume no genotype-environment interactions. It appears that we also assumed that the additive and dominance effects are uncorrelated because there is no covariance term for these factors. However, the least-squares method used to find the additive effects makes the additive and dominance contributions uncorrelated.

Table 11.7. All possible ways to make a genotype.

Genotype	*Frequency*	X_m	X_p	X_{mp}
A_1A_1	p^2	α_1	α_1	$a_{11} - 2\alpha_1$
A_1A_2	pq	α_1	α_2	$a_{12} - \alpha_1 - \alpha_2$
A_2A_1	pq	α_2	α_1	$a_{12} - \alpha_2 - \alpha_1$
A_2A_2	q^2	α_2	α_2	$a_{22} - 2\alpha_2$

Up to this point, both of our models assume that the genetic contribution to a trait is due to a single locus. Obviously, most quantitative traits will be affected by more than one locus. The generalization to many loci for di-allelic loci comes from simply rewriting Equation 24 as

$$P = \sum_i [X_m(i) + X_p(i) + X_{mp}(i)] + \varepsilon, \qquad ...(25)$$

where $X_m(i)$, $X_p(i)$, and $X_{mp}(i)$ are the genetic contributions from the *i*th locus and the sum is over all loci that affect the trait. The variance in the phenotype is

$$V_P = V_A + V_D + V_E,$$

where now the additive variance is defined as the sum of the additive variances of the individual loci,

$$V_A = \sum_i V_A(i).$$

The dominance variance is summed across loci as well. Once again, we have slipped in an assumption to remove covariance terms. This time, we assume that there is no interaction between alleles at different loci. If there were an interaction, the additional variance due to the interaction would be called the ***epistatic variance***, V_I, and the phenotypic variance would be

$$V_P = V_A + V_D + V_I + V_E.$$

In this more general setting, the heritability is still defined as

$$h^2 = \frac{V_A}{V_P}$$

and is often called the ***narrow sense heritability***. The broad sense heritability is the ratio of all of the genetic variances to the phenotypic variance,

$$H^2 = \frac{V_A + V_D + V_I}{V_P}.$$

The covariance between relatives with many noninteracting loci is

$$\mathrm{Cov}\{P_X, P_Y\} = rV_A + r2V_D,$$

where, as with the variance partitions, the additive and dominance variances are the sums of the variances of the individual loci. Thus, as far as variances and covariances go, the one-locus and multilocus models give the same results.

There is one very important property of the two-allele model that does change when more loci are added: The distribution of the genetic effects approaches a normal distribution. Notice that the distribution of height approximates that of a normal distribution, albeit crudely because of the small sample size. There is no a priori reason why the phenotypic distribution should be so, well, normal. The Central Limit Theorem from probability theory does provide a partial explanation. This theorem states that the distribution of the sum of independent random variables, suitably scaled, approaches a normal distribution as the number of elements in the sum increases. The approach to the normal distribution is usually quite fast. For example, the distribution of the sum of as few as 10 uniform random variables looks remarkably like a normal distribution. Thus, if as few as 10 loci contributed to a trait, the genetic contribution would look normal.

Of course, the phenotypic distribution also has an environmental component that must itself be approximately normally distributed if the phenotypic distribution is to be normally distributed. In fact, this appears to be generally true as judged from an examination of the phenotypic distribution of individuals that are genetically identical, as occurs, for example, in inbred lines. Perhaps the environmental component is also the sum of many small random effects that add to produce their effects on the phenotype.

One of the outstanding problems of quantitative genetics concerns the number of loci that contribute to the variation in a trait. Speculation has ranged from a very large number of loci, each with a very small and roughly equivalent effect, to very few loci, each with a relatively large effect. Direct genetic analysis of some traits (e.g., the number of bristles on a *Drosophila*), supports the latter view. With modern molecular techniques, this issue should be resolved in the near future.

Intensity of Selection

This short section provides a quick derivation of the intensity of selection. The derivation uses properties of the normal distribution.

The probability density of parental phenotypes is the normal density with mean zero and variance V_P,

$$\frac{1}{\sqrt{2\pi V_P}}\exp\left(\frac{-x^2}{2V_P}\right).$$

Thus, the proportion that is selected is the area

$$p = \int_\alpha^\infty \frac{1}{\sqrt{2\pi V_P}}\exp\left(\frac{-x^2}{2V_P}\right)dx,$$

where a is the largest number that is less than the values of the selected parents. The proportion may be simplified by changing the variable of integration to y $= x/\sqrt{V_P}$,

$$p = \int_{\alpha/\sqrt{V_P}}^\infty \frac{1}{\sqrt{2\pi}}e^{-y^2/2}dy. \qquad ...(26)$$

The selection differential is the mean of the selected parents, which is the mean of the parents. The density of the selected parents is the truncated normal density

$$\frac{C}{\sqrt{2\pi V_P}}\exp\left(\frac{-x^2}{2V_P}\right), x > \alpha,$$

where the constant C is chosen to make the integral of the density equal to one,

$$C = \left(\int_\alpha^\infty \frac{1}{\sqrt{2\pi V_P}}\exp\left(\frac{-x^2}{2V_P}\right)dx\right)^{-1}$$

The mean of the selected parents is thus

$$S = C\int_\alpha^\infty \frac{x}{\sqrt{2\pi V_P}}\exp\left(\frac{-x^2}{2V_P}\right)dx.$$

As with the proportion selected, the selection differential may be simplified by changing the variable of integration to y = $x/\sqrt{V_P}$,

$$S = \sqrt{V_P}\,i(\alpha/\sqrt{V_P}), \qquad ...(27)$$

where

$$i(\alpha/\sqrt{V_P}) = \frac{\int_{\alpha/\sqrt{V_P}}^\infty ye^{-y^2/2}dy}{\int_{\alpha/\sqrt{V_P}}^\infty e^{-y^2/2}dy} \qquad ...(28)$$

is called the *intensity* of selection.

Notice that the proportion selected, as given by Equation 26, and the intensity of selection, as given by Equation 28, are both functions of the quantity $\alpha/\sqrt{V_P}$. In fact, they are both monotonic functions of $\alpha/\sqrt{V_P}$ is a decreasing function and S is an increasing function. As a consequence, for each value of p there is a unique corresponding value of the intensity of selection. Thus, we can view the intensity of selection as a function of p rather than as a function of $\alpha/\sqrt{V_P}$. Unfortunately, the functional relationship cannot be written as a simple formula because both p and $i(\alpha/\sqrt{V_P})$ are functions of integrals. However, it is easy to evaluate the integrals with a computer and in so doing. This figure relates the intensity of selection to the proportion of selected individuals. The selection differential is then obtained using Equation 27, which should now be written as

$$S = \sqrt{V_P}\, i(p),$$

which is the same as Equation 12.

The approach used here to obtain the selection differential uses an implicit assumption that there are a large number of measured individuals that are used to select the parents. If the number of measured individuals is 20 or fewer, then a more accurate intensity-of-selection function is required. Such functions may be found in any standard text on quantitative genetics.

INDEX